Production Ergonomics:
Designing Work Systems to Support Optimal Human Performance

Cecilia Berlin, PhD & Caroline Adams, MEng

]u[

ubiquity press
London

Published by
Ubiquity Press Ltd.
6 Windmill Street
London W1T 2JB
www.ubiquitypress.com

First published 2017

Cover illustration by Camia Pia Pettersson
Cover design by Amber MacKay

Printed in the UK by Lightning Source Ltd.
Print and digital versions typeset by Siliconchips Services Ltd.

ISBN (Paperback): 978-1-911529-12-5
ISBN (PDF): 978-1-911529-13-2
ISBN (EPUB): 978-1-911529-14-9
ISBN (Mobi): 978-1-911529-15-6

DOI: https://doi.org/10.5334/bbe

The full text of this book has been peer-reviewed to ensure high academic
standards. For full review policies, see http://www.ubiquitypress.com/

Suggested citation:
Berlin, C and Adams C 2017 *Production Ergonomics: Designing Work Systems to Support
Optimal Human Performance*. London: Ubiquity Press. DOI: https://doi.org/10.5334/bbe.
License: CC-BY 4.0

To read the free, open access version of this book online, visit
https://doi.org/10.5334/bbe or scan this QR code with your mobile
device:

Contents

Acknowledgements

The authors would like to thank:

- Professor Rikard Söderberg, director of the Area of Advance PRODUCTION at Chalmers University of Technology, for funding this book.
- Chalmers University of Technology's Education Quality Improvement initiative, for funding the publishing of this book.
- All the students in the production ergonomics and work design classes of 2013, 2014, 2015 and 2016 for inspiring this book and alerting us to typos, unclear wordings and suggesting additions to the book.
- Caroline Dedering and Guðbjörg Rist Jónsdóttir for excellent support in researching the field of social sustainability.
- Ann-Christine Falck, for sharing her knowledge, passion and experience of ergonomics and economics for the benefit of this book.
- Caroline Dedering, Sandra Mattsson, Jonatan Berglund, Lars Medbo, Anna Dean and Maja Bärring for proofreading sections of the book and providing feedback.
- Associate professor Peter Almström at Chalmers University of Technology for being a great colleague and co-instructor in the course "Production Ergonomics and Work Design", and for feedback on this book's contents.
- Thanks to our anonymous reviewers who during the publishing process supplied helpful advice that helped to shape this book towards its final version.
- Thanks to all credited individuals, groups and organizations that kindly granted us permission to use their images as illustrations.
- Massive thanks to the staff at Ubiquity Press for making this book better! We particularly thank our editor Samuel Moore for consistent, kind encouragement and a great deal of unwavering patience during the publishing process, and our copy editor Rebecca Mosher for sharpening the language and clarity of the text.

Finally, Cecilia would like to thank Caroline for jumping into the process of writing and structuring this book with much-appreciated skill, dedication, humour and patience.

Gothenburg, Sweden
December 29, 2016

Preface

Hello, and welcome to the wonderfully complex world of production ergonomics. This book is meant to introduce engineering students, particularly in the area of production engineering, to the huge potential of designing better industrial workplaces on the basis of a solid foundation of knowledge in *ergonomics* (the scientific study of human work), also known as *human factors*. We have aimed to do this in a way that is quickly accessible, comprehensive, and explained at various levels of detail depending on the engineer's future working role. In a teaching context, this book is best used as a reference companion alongside analytical assignments, case studies or a practical workplace improvement project, where students are tasked with analysing the improvement potential of a workplace and then designing a solution. Using this book, we hope that we have made it easier for the reader to design workplaces that live up to various ergonomic "best practices" and guidelines for efficient, safe, healthy and effective work.

The book started as a course compendium at the Master-level course "Production Ergonomics and Work Design" at Chalmers University of Technology in Gothenburg, Sweden. It was tailored to the curriculum of that course and owes much of its structure and contents to 1) the practical, project-based needs of its students, and 2) the Swedish/European context. While developing the course compendium into a book, we have done our utmost to increase the international perspective and to alert the reader to the different roles which may find different aspects of workplace design particularly valuable for their day-to-day job. We also hope that this will help the reader realize that workplace design is a team effort and that the contribution of different roles are needed to build and maintain a well-functioning, healthy and coherent work system.

Designing, building and evaluating production workplaces is a complex skill set that requires a gradual acquisition of knowledge about human needs and prerequisites, critical and creative thinking and methods, and knowledge about societal drivers that shape future demands on production workplaces. Letting these skills mature together and inform each other is what separates a masterful, proactive workplace designer from one who is limited to checklists and "fire-fighting". The authors hope that this book will help to show how each of the areas in it are interconnected, with the human worker's capability, limitations and requirements at its nexus.

Humans are able to perform fantastic feats when they are prepared, supported, content, trained, focused and at their physical and mental best. At other times, they may also be limited in their performance because they are fatigued, bored, injured, confused, discontent, physically weak, demotivated, elderly, beginners – the list of considerations is long. So to make future workplaces more robust, it is necessary for an engineer to learn how to design for the range of how very different human workers' needs and abilities can be. We hope that our readers will realize that some human performance aspects are so nuanced and dynamic that building flexibility into your system to support individual variation becomes a good investment.

The primary audience of this book is budding workplace designers – particularly engineering students who may someday be responsible for the design, work organization and layout of factory-level production environments. For that audience, the book tries to cover basic knowledge of human needs, including physical, cognitive and social prerequisites for performing work, and then moves on to methods for successfully implementing, running and evaluating work systems. For the benefit of this audience, the book is organized into two parts in order to approach the subject from an "inside looking out" perspective. *Part 1 – Understanding the Human in the System*, starts with knowledge of the individual human's capabilities and prerequisites at work; moving on to interaction with technology, cognitive tasks and other humans. *Part 2 – Engineering the System around Humans* moves on to analysis and design methods, tools and skills, and zooms out to macro perspectives of economy, society and social sustainability. Just to keep reminding ourselves of the relevance to engineers, each chapter begins with a short reflection called "Why do I need to know this as an engineer?", to describe how that knowledge may be valuable and help the workplace designer avoid pitfalls of missing something in their design considerations.

In some chapters, we go into great detail in order for our reader to learn and exercise specific skills. At other times, the explanations are more aimed at giving you an overview, so that you can fruitfully begin to seek further knowledge on your own and discuss your work with experts on related subjects. If you want to go into depth, you can find more to read in the references and bibliographies at the end of each chapter. However, at all times we have strived to keep the language accessible and intentionally less academic than some of the research materials it builds upon. We have also introduced some different professional roles that engineers may end up in once they start working in an organization – we use these roles at the beginning of each topic chapter as a filter for our reader to understand which topics are central to different stakeholders.

All in all, this book aims to provide you with a good mix of theory, methods, design checklists, large-scale perspectives, stakeholder perspectives, resources for further reading and examples – things you will need in your arsenal when convincing other stakeholders that your design proposal is a feasible socially and economically sustainable idea, both in the short and long term. Our hope is that this book will provide you with a good ladder up to understanding the human being's strengths and limitations, so that you can design a robust, high-performing, economically responsible system. What you will learn is that taking care of the humans in your work system is a gift that keeps on giving.

We hope you will enjoy this book!

The Authors

Cecilia Berlin, PhD

Caroline Adams, MEng

CHAPTER 1

Introduction

THIS CHAPTER PROVIDES:

- An introduction to the different roles in production engineering that may need to concern themselves with ergonomics/human factors knowledge.

How to cite this book chapter:
Berlin, C and Adams C 2017 *Production Ergonomics: Designing Work Systems to Support Optimal Human Performance.* Pp. 1–12. London: Ubiquity Press. DOI: https://doi.org/10.5334/bbe.a. License: CC-BY 4.0

- An overview of the wide variety of aspects covered under the umbrella term ergonomics/human factors.
- A brief discussion on the relevance of production ergonomics to the performance of a production system.
- A history lesson of how ergonomics/human factors developed.
- An overview of the contents of this book and how they are organized.

WHY DO I NEED TO KNOW THIS AS AN ENGINEER?

Sometimes, a bit of history goes a long way to explain why certain things in a discipline are considered important. For an engineer, it may be good to know what the starting point was before any real thought was put into methodically improving the human aspects of production work. It will also help you understand why ergonomics covers so many areas and is such a diverse and complex discipline.

The discipline of ergonomics is not nearly as old as medicine, or even industrialization, but it arose as a consequence of an extreme social situation: World War II. With all the able-bodied young men drafted to war, industrialists faced a need to suddenly adapt workplaces to the needs and limitations of a new, more diverse workforce consisting of women, physically disabled, and other previously overlooked groups of society. At the end of the war, society itself had changed to the point where it was acceptable for many of these groups to remain in employment.

While this first effort concerned itself mostly with physical work, later historical developments showed that it was possible to also improve workplaces in relation to human mental capability, teamwork and organizations. Today, it is in the best interests of most industries to build workplaces where the greatest possible diversity of people are able to perform well, meaning that physical, cognitive and organizational sides of ergonomics are equally powerful aspects in the design of inclusive workplaces.

As ergonomics widened its scope, it became the concern of more and more stakeholders. Today, it is worthwhile to know that ergonomics has the potential to concern, engage and/or provoke many more people than just the workplace designer, the ergonomist or the worker.

1.1. What is ergonomics/human factors?

For many people, the word *ergonomic* is associated primarily with comfy office chairs, the correct height of computer screens, computer mice and consumer products that have been (sometimes randomly) labelled "ergonomic", like kitchenware, backpacks or gardening tools. The word itself comes from the Greek roots *ergon* (work) and *nomos* (laws) and roughly translates to "the science of work", focusing on human activity.

> ### *Ergonomics*
>
> from the Greek words Ἔργον [ergon = work], and Νόμος [nomos = natural laws]; "the science of work"

But *ergonomics* (or *human factors,* an equivalent term used more commonly in North America) in general is a very wide term. Ergonomics can signify anything from the physical activities and demands of the job, to how the human mind understands instructions and interfaces, to how work organization, teamwork and motivation influences human well-being and efficiency. Furthermore, it may include aspects of aging, working in extreme environments (such as fire fighting, working in freezer rooms or mines), working with protective gear (such as protection gloves, heavy jackets, helmets, etc.). In short, almost any aspect of work involving human activity can be approached from an HFE (Human Factors and Ergonomics) perspective.

Simply visiting the Human Factors and Ergonomics Society (2015) website reveals that they are organized into as many as 23 different "technical groups" which specialise in applying ergonomics knowledge and practice to the areas in Table 1.1.

1.2. The purpose of production ergonomics

It can be assumed that anyone in charge of a production system would want all of its sub-components to function together with as much ease and efficiency as possible. When part of that production system is human, the performance of the system as a whole may vary depending on the daily form of the human workers. Although humans have great potential to bring flexibility, innovation and problem-solving skills to the production system, they are at risk for developing work-related musculo-skeletal disorders (alternately abbreviated MSDs or WMSDs) as a result of physical work that overloads the human body. Symptoms of such risks include discomfort, pain and recurring

Table 1.1: The 23 technical groups of the Human Factors and Ergonomics Society as of 2015.

• Aerospace Systems	• Individual Differences in Performance
• Aging	• Internet
• Augmented Cognition	• Macroergonomics
• Cognitive Engineering and Decision Making	• Occupational Ergonomics
• Communications	• Perception and Performance
• Computer Systems	• Product Design
• Education	• Safety
• Environmental Design	• Surface Transportation
• Forensics	• System Development
• Health Care	• Test and Evaluation
• Human Performance Modelling	• Training
	• Virtual Environments

injuries, and the consequences of unhealthy loading include suffering, inability to work and high costs for the company (in terms of compensation, productivity losses and replacement of personnel). Also, human mental capacities are dependent on sufficient support, stimulation and opportunities for rest. Without these health factors, confusion, irritation, misinterpretation and serious errors can occur, potentially causing material or personal harm. Finally, the interactions between human workers can at best be a source of support, stimulation and a feeling of identity, but if they are dysfunctional they can also cause demotivation, dissatisfaction and lack of engagement. In other words, the purpose of production ergonomics is to design a workplace that is proactively built to remove the risks of injury, pain, discomfort, demotivation and confusion.

How a company chooses to handle production ergonomics may vary with their size, organizational form, previous history of involving ergonomics expertise, project experiences, access to standards, previous knowledge of methods and tools, and expectations of different stakeholders in the company on the person put in charge of ergonomics. A proactive approach towards production ergonomics is characterized by getting ergonomics knowledge into the early planning stages, seeing ergonomics as a source of long-term cost savings and a high regard for keeping the workforce healthy. A reactive approach, on the other hand, usually leaves ergonomics issues and risks unaddressed until problems start cropping up, such as worker pain, injuries and sick leave. Quite frequently, companies with a reactive ergonomics approach will try to solve problems with a healthcare service angle, which only serves to take care of the symptoms and not the root cause of the problem, which then remains as a risk to other workers.

1.3.   Historical development of ergonomics and human factors

The modern history of ergonomics in the Western world dates back to the 1940s, during World War II. As a result of the demands of warfare, many able-bodied young men were drafted to participate in the war effort, leaving their civilian work (e.g. in factories). At the same time the war effort demanded new military vehicles, equipment and instruments, giving rise to a new form of industry, which needed to produce products at a high pace with high quality, and therefore required more manpower. This meant that production on the home front needed to be staffed by the population who remained. The shift included re-training and transferring male workers from civilian businesses to the warfare industry, but also called on women, the elderly, disabled and previously excluded social groups to fill the demand. Recruitment efforts resulted in a new form of state propaganda that gently challenged societal norms, such as by stating that women should be capable of performing assembly jobs as it was not completely different from high-precision housework. As a result of this drastic diversification of the working population, industries began investing in physical aids (such as new tools and devices for lifting and supporting heavy machinery) to enable the presumably weaker workers to carry out assembly jobs at a maximum level of efficiency and productivity.

This first shift of the 1940s, where industrial attention was focused on the human functioning in a technical system, is referred to as the "physical generation" of ergonomics developments. The focus was on physical characteristics of the human body, anthropometry, posture, health and safety, perceptual capabilities, and how they affected the design of technology. Scientific and practical developments have since continued in the field of physical ergonomics to the present day, with plenty of influence coming from sports medicine (emphasizing physical performance) and medical monitoring of health (using measurement instruments such as electromyography, EMG, to study human muscle use).

About 20 years later, in the 1960s, scientific developments were made in the area of computers and robotics, which presented many new possibilities but were also perceived by some as a threat to the human worker; would robots take over all human jobs? Would they, indeed, take over the world? While these fears were left hanging, science and engineering underwent a change of perspective; instead of looking at how human needs influenced technology, the demands of technology on humans were highlighted instead, leading to a focus on cognitive psychology, mental workload (and overload), skill, cognitive limitations (e.g. memory) and psychological factors during work. The 1960s brought with them a rapid development of computer interfaces and control rooms.

Yet another 20 years later, in the 1980s, HFE researchers began to realize that in spite of their extensive knowledge in the areas of physical and cognitive ergonomics (uniting the body and the mind), it was seldom that that knowledge was allowed to influence the design of workplaces and machinery. They realized that there was a strong dependency between technology and organizations, and that the effect of interpersonal relationships that influence design outcomes was greater than previously thought. This led to a view of ergonomics work being part of a "sociotechnical system" with greater focus on the context and the stakeholders surrounding ergonomics, leading to the third generation known as "the Macroeconomic generation". Sometimes also referred to as "organizational ergonomics", this branch explores the role of ergonomics within an organizational context with multiple stakeholders with different agendas. It also addresses the fact that working successfully with ergonomics is a balance of considerations; this is especially true for production ergonomics, where the goals of production engineers, economists, managers, human factors professionals and operators can all influence decisions and changes in workplace improvement.

Dray (1985) describes this historical development as the "three generations of ergonomics". However, the evolvement of HFE did not stop in the 1980s. Yet another 20 years onward, in the year 2000, the council of the International Ergonomics Association (IEA) decided to strengthen the industrial relevance of ergonomics by declaring globally that ergonomics was not only focused on the human's well-being, but also on the efficiency, performance and productivity of work systems and machines. There was also a need to signal equality between the terms *ergonomics* and *human factors*, as both terms were used to signify similar concerns, but with some variation both between countries and industrial sectors (for example, Scandinavian countries and the manual assembly industry have a tendency to use the term *ergonomics*, while the term *human factors* is more predominant in North America and in the nuclear industry). Therefore, the association issued the following definition:

Definition of the International Ergonomics Association, IEA (2000):

"Ergonomics (or human factors) is the scientific discipline concerned with the understanding of interactions among humans and other elements of a system, and the profession that applies theory, principles, data and methods to design in order to optimize human well-being and overall system performance."

This definition remains the official one for ergonomics and human factors, but the IEA recognizes the physical, cognitive and organizational branches as the three main "domains of specialization".

Modern developments, primarily from the 1990s and onwards, have seen an increase in ergonomics simulation, i.e. the introduction of ergonomics analysis tools into 3-D computer design environments. Specific software has been developed to enable the simulation of work positions and work-related actions in a 3D CAD environment, using a human form called a manikin. Manikins of both genders can be scaled to different sizes in order to investigate whether the extremes of the human population will be able to work in a proposed environment without exposing themselves to physical risk for injury. This type of software is predominantly found in technologically mature, economically profitable industrial sectors producing large, complex products, notably the automotive industry.

Another recent development which has gained popularity over the past decade is an increased emphasis on the effects of aging; demographic developments in the Western world suggest that it will be necessary to keep production employees in the workforce for a longer working life, since an outflux of retirees would cause industries a lot of brain-drain, or loss of know-how and competence. This will pose challenges in terms of designing and adapting the workplace to the changed prerequisites and demands of the human body as it ages, while at the same time supporting the worker in performing their job without loss of precision, productivity or efficiency. At the same time, workplaces must be designed to attract and support a new generation of workers, who will most likely be required to perform increasingly complex jobs from the beginning of their working lives. Today, this combination of challenges has notably gained attention from governments and the academic world since the 2010s, resulting in an increased focus on placing social sustainability alongside economic and environmental sustainability.

1.4. How are ergonomics and human factors connected to engineering?

Engineers have a distinct advantage as workplace designers and improvers: companies that hire engineers expect them to independently come up with analyses and suggestions for change as part of improving systems and operations. Expectations from company leadership on an engineer's mind-set and skills often lead to a role where they are trusted to come up with practical suggestions and even make decisions that change the workplace.

Other roles with ergonomics and human factors knowledge, such as ergonomists, occupational health and safety (OHS) agents, medical/ health service staff, consultants, etc. may not always have the same mandate, expectation or training to suggest design changes, purchases, work task modifications, etc. – and if they do, those with a medical or physiotherapeutic background may be limited in scope to merely providing an analysis output, but not to contribute towards a new design solution (unless the company in question is ergonomically mature enough to make this possible using cross-functional teams; but this practice is not to be taken for granted). Also, a disadvantage of addressing ergonomics and human factors from the medical/ health angle is that they are often not able to act until workers have actually been complaining or have gotten injured – and in such cases, interventions may end up tailored to easing the situation only for the injured worker on an individual basis. It may be hard from that angle to argue for any comprehensive changes in a proactive manner, if management is not convinced that the problem can recur and cause trouble again. Therefore, workplace change agents with an engineering role have a greater leverage to make sustainable improvements, because they may be able to do something to address the root cause in the work system that may be a risk for many workers. In other words, an engineer who has good knowledge of ergonomics (and its monetary value) can have a very positive long-term impact on business because their knowledge about human needs and capabilities can be translated into feasible system design changes that can avert systemic health and safety risks. That is, engineers can do this, *if* they are educated and trained to recognize matters of human well-being and system performance as part of their work to make a workplace more efficient, productive and socially and economically sustainable.

1.5. What's in this book?

Preface	This section explains how and why this book was written. Reading it may result in understanding the authors' intentions better.
1. Introduction	In this chapter, we introduce background knowledge of the ergonomics/ human factors domain and how it relates to production engineering.
PART 1 – Understanding the Human in the System	
2. Basic Anatomy and Physiology	The human body is amazing in many ways, but its needs, abilities and limitations change over time. Getting to know how it responds both in sickness and in health is a good basis for doing engineering work to support and save it.
3. Physical Loading	Here, the basic knowledge we have of anatomy and physiology is combined with principles of classical mechanics to translate it into engineering terms.
4. Anthropometry	Designing a workplace is something you do for more than one person to include as many potential users as possible – this chapter helps you figure out how to design for populations, rather than just a few people.
5. Cognitive Ergonomics	Here we devote our attentions to the human mind and senses, and gain an understanding of the needs and limitations that affect our ability to understand information and take action.
6. Psychosocial Factors and Worker Involvement	The human does not operate alone, but is influenced by interactions with others and has needs and limits for how that interaction should take place. Here, we examine workplace health factors having to do with organization, support, stress, mandate and freedom to act and influence the workplace.
PART 2 – Engineering the System around Humans	
7. Data Collection and Task Analysis	This chapter introduces data collection for the purpose of improving workplaces, and the basics of Task Analysis in order to structure the engineer's understanding of intended and/or existing operations.

8. Ergonomics Evaluation Methods	This chapter introduces methods for determining if there are ergonomic injury risks present in an existing (or simulated) workplace.
9. Digital Human Modeling	This chapter briefly introduces how ergonomics simulation can be used to test a workplace in a CAD (computer-aided design) environment at early stages of development without the need for costly mock-ups and costly materials.
10. Manual Materials Handling	This chapter explains some different ways that the handling, moving and storage of material significantly affects ergonomics and productivity.
11. The Economics of Ergonomics	It is often not enough to know why and how ergonomics is good for the human body and its abilities – quite often, improved work conditions mean better productivity and economic returns. Knowing how can make you skilled at persuading management to invest in workplace improvements.
12. Environmental Factors	Here, the effects on our well-being and performance that stem from our environmental surroundings are described, as well as the concept of "comfort zones" for creating optimal work conditions.
13. Social Sustainability	Here, the long-term impacts of what you can do as a work designer are explained. Sustainability has to do with making long-lived, healthy, competitive workplaces that contribute to creativity and innovation.
Notes for Teachers	This section is aimed at instructors and explains the wider perspectives of using this book as part of an engineering curriculum.
PART 3 – Workplace Design Guidelines	
This section contains a compilation of design guidelines for the different topics covered in the book.	

1.6. Different engineering roles act on different types of knowledge

Engineers may end up playing a variety of different (sometimes overlapping) roles in their professional career, each with their distinct scope, system level and operational concerns – some switch between several of these throughout their working life, depending on how specialized their working role is and at what system level they are expected to address problem solving. For example, an engineer may act on a specialized, operative level with responsibility for a single production line, which would require specific methods and knowledge to optimize for human well-being and performance. Other engineers end up at a management level, where they are perhaps not served by anatomical knowledge and ergonomics evaluation methods, but may impact it greatly by having responsibilities for economics, personnel well-being and approving investments in new equipment. Yet others may act in a more visionary way to orchestrate a production system on a macro scale, involving supply chain operations and a sustainability vision.

At any one of these levels, knowledge of ergonomics and human factors can be a vital part of continuous improvement work, as well as a sound business practice where the value of healthy, knowledgeable and motivated workers is proactively supported and preserved before any problems or system inefficiencies arise, thanks to the engineer understanding what is required of a system for its human components to perform at their best.

Since this book aims to give both detailed knowledge about the human body and mind's capabilities and prerequisites (Part 1) as well as to provide actionable ways to design and improve work systems (Part 2), we have identified some different engineering roles (see Figure 1.1) that may be useful as "filters" to sift through the knowledge in this book, both while studying (if you have a future work role in mind) and later in life as a practicing professional. For the latter group, we hope that the book can continue to serve as a handy reference for making prioritizations, business cases and design decisions. It may also be helpful to be aware of the perspectives of other actors in a production organization, as they may require a tailored set of arguments to become convinced of the benefits of a workplace change initiative.

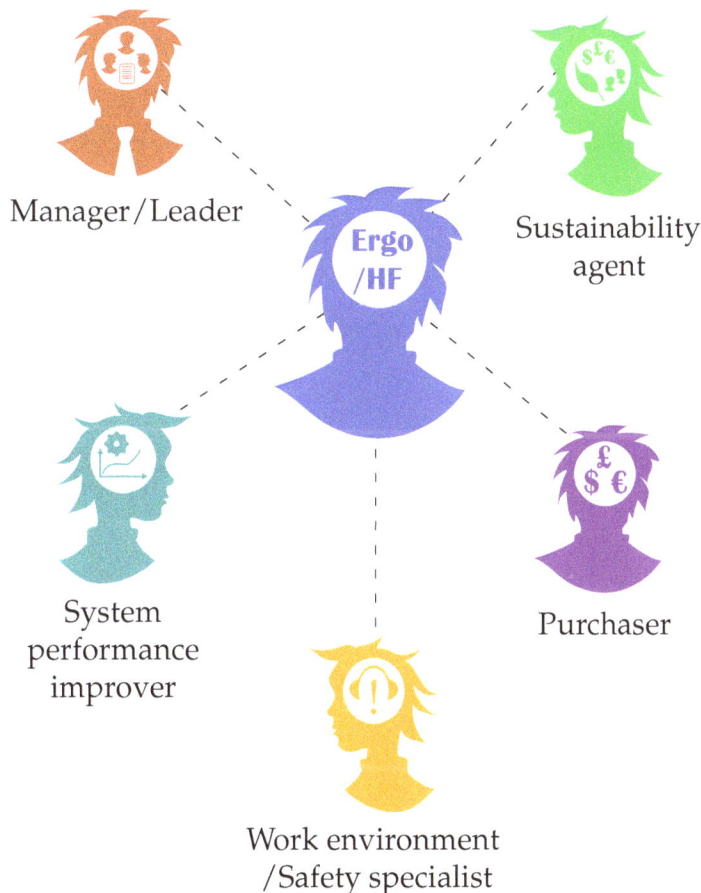

Figure 1.1: Working roles in which an engineer can use ergonomics and human factors knowledge to positively impact a workplace.

Illustration by C. Berlin.

The manager / leader	This person has a wide scope of responsibility in a company, addressing aspects like: the recruitment, training, performance and well-being of employees; having the mandate of whether to approve improvement projects and make investments; running a productive and feasible business, where employees are treated as a valuable asset; and aligning operations with an overall organizational vision, such as a sustainability strategy. This person needs a macro-system view, an understanding of conditions that support worker well-being on an individual and team level, and economical aspects of work system performance.
The system performance improver	This person is responsible for the performance and improvement of a particular system or sub-system (for example, the efficiency of a production line) and acts independently to make a current-state analysis, which in turn acts as a basis for suggesting improvements. This role must understand the economic gains of good ergonomics to make a compelling business case for changes, and relies on data collection, ergonomics evaluation methods and tools, and an understanding of which conditions allow humans to perform physical and mental work well.
The work environment / Safety specialist	This person has a particular focus on the workers' well-being and safety. This means that a solid knowledge of the capabilities and limitations of the human body and mind at work is essential for this role, in order to avoid harmful loading, distraction and repetitive strain. This person also needs to understand how work environmental factors influence human performance, and need to be able to use guidelines and standards to ensure the design of safe high-performance work environments.
The purchaser	Although perhaps not the most typical engineering role, this one has a considerable say in whether an improvement is made possible or not (and may overlap with other roles). When this person has an understanding for the type of investments that lead to an economically sustainable work environment with few worker ill-health issues, then money can be used wisely to invest in solutions with a synergetic systems perspective (rather than a reactive, individual-based one) that will have a lasting beneficial impact. They often need to consider legislative demands and time-horizons for expected payback on an investment.
The sustainability agent	Finally, an increasing concern for many organizations is that of sustainability in all business aspects; this means balancing social, economic and environmental aspects in order to ensure that continued operations will have a positive impact on people, planet and profit. But how is this connected to ergonomics and human factors? We argue that sustainability – particularly social and economical aspects – can be addressed both in a global macro-perspective and a local, company-level perspective, and that with a solid understanding of human needs and how they translate into requirements on a workplace, engineers who design and improve workplaces can contribute to more socially and economically sustainable production systems.

At the beginning of each topic chapter (Chapters 2, 3, 4, 5, 6, 7, 8, 9, 10, 11, 12 and 13), a short statement is given explaining how particular roles can use the knowledge in each chapter to have a beneficial impact on worker well-being and system performance, including the perspective of good business sense.

1.7. References

Dray, S. M. (1985). Macroergonomics in Organizations: An Introduction. In: Brown, I.D., Goldsmith, R., Coombs, K. and Sinclair, M. (Eds.) *Ergonomics International,* 85: 520–525. Taylor and Francis, London.

Hendrick, H. & Kleiner, B. (2001). *Macroergonomics: An Introduction to Work System Design.* Santa Monica, CA: Human Factors and Ergonomics Society. Design, Human Factors & Ergonomics Society. ISBN 0-945289-14-6

Human Factors and Ergonomics Society. (2015). Technical Groups. [Online] Available from: https://www.hfes.org//Web/TechnicalGroups/descriptions.html [Accessed 1 Oct 2015].

IEA, International Ergonomics Association. (2000). Definition and Domains of ergonomics. [Online] Available from: http://www.iea.cc/whats/index.html [Accessed 17 Jan 2014].

PART I

Understanding the Human
in the System

CHAPTER 2

Basic Anatomy and Physiology

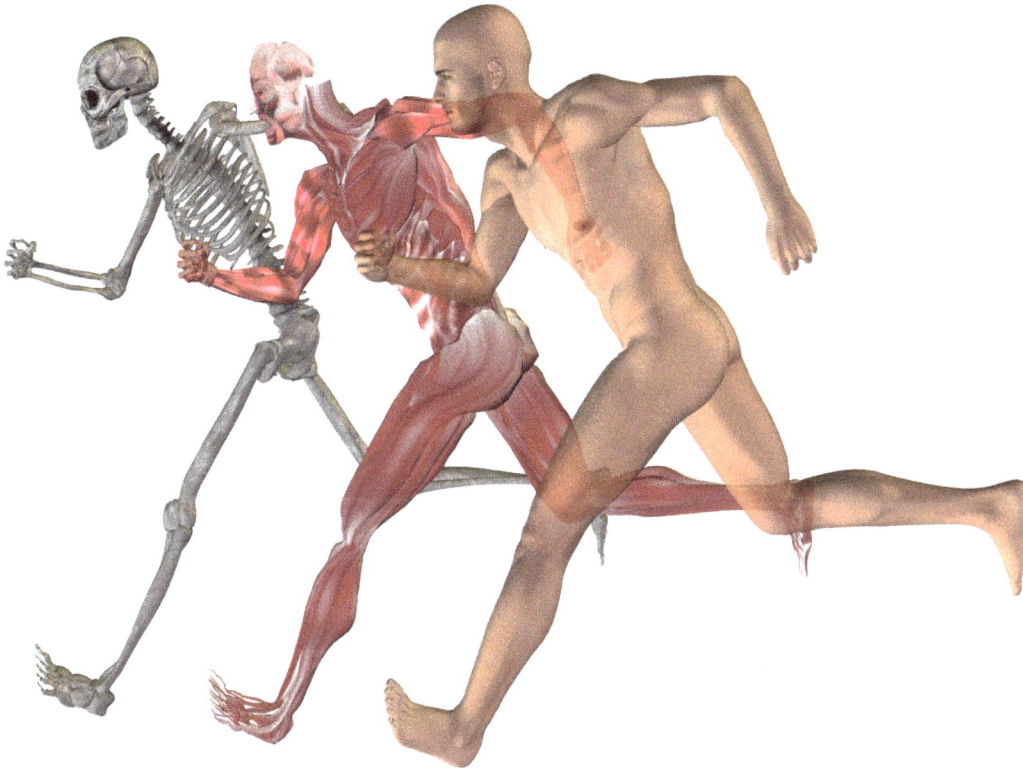

How to cite this book chapter:
Berlin, C and Adams C 2017 *Production Ergonomics: Designing Work Systems to Support Optimal Human Performance.* Pp. 15–48. London: Ubiquity Press. DOI: https://doi.org/10.5334/bbe.b. License: CC-BY 4.0

THIS CHAPTER PROVIDES:

- Descriptions of how the different structures of the musculo-skeletal system are shaped (anatomy) and how they work and respond to loading (physiology).
- A description of current injury statistics regarding musculo-skeletal disorders and how big the problem is for production industry.

WHY DO I NEED TO KNOW THIS AS AN ENGINEER?

From a physical point of view, having a basic understanding of the human body's strengths, abilities, and limitations is an important basis for making well thought-out tweaks to the design of the workplace, in order to build work systems that are not a risk to human health or performance. Knowing how your muscles, bones and joints work may seem like a far cry from your engineering work, but it will significantly help your understanding in later chapters where physical loading and methods of ergonomics evaluation are discussed. Another thing this chapter does, is to provide a *limited* description of anatomy and physiology; it will not go into as much detail as an anatomy book, but provides the level of detail needed to understand some of the methods that will be explained later.

If you as an engineer start using ergonomics evaluation methods without first gaining the knowledge in this chapter, the reasoning that those methods are based on would probably remain a mystery. You could still use them, but if you were questioned about their limitations or why you were using them, you would probably not be able to explain their validity, or reason about unexpected results. Knowing about the human body and its strongest and weakest positions can also encourage even an engineer to adopt more healthy movement, posture, loading and sitting behaviours in their everyday life – and *that* awareness is the best basis for becoming a great workplace designer.

WHICH ROLES BENEFIT FROM THIS KNOWLEDGE?

The engineer who acts as *system performance improver* or *work environment/ safety specialist* is likely to observe and analyze actual physical work being performed, before making a recommendation or a design proposal. With knowledge of how the human body functions optimally and how it is limited in strength, stamina and injury recovery, the engineer can avoid building potential risks for MSDs (musculo-skeletal disorders) into the work system. These roles may also interact with workers who complain in an imprecise manner about pain or discomfort, or with medical or health personnel who are not trained in using ergonomics evaluation methods to evaluate risks. With basic knowledge of anatomy and physiology, the engineer can communicate effectively with these stakeholders about risks and possible solutions.

2.1. Musculo-skeletal disorders

Our ability to work – in any way – is completely dependent on our physical health. When we feel unease, discomfort, pain or numbness, we may be able to ignore the body's warning signals and still perform work, but the body will perform slower; with less power, quality and precision; with more errors; and at worst, resulting in serious accidents. A very real problem that is faced by all production industry is when the limit has been passed for what a human body can tolerate, resulting in a worker needing to go on *sick leave,* i.e. be absent from work to recover from physical disability. If the disability affects the worker's physical ability to move and handle loading, then the worker is said to be suffering from a *work-related musculo-skeletal disorder* (abbreviated either as WMSD[1] or just MSD). MSDs are defined as a heterogeneous group of disorders caused by a multitude of potential (physical) factors. Pain, discomfort and fatigue are considered common first symptoms, while more obvious signs of the presence of an MSD include loss of function, limited movement range and loss of muscle power.

The costs of a worker taking sick leave can balloon to huge proportions: not only does the employer in many cases need to cover the worker's sick leave compensation and rehabilitation costs, but there are also the costs of recruitment, training of new personnel and losses of productivity and quality until a new employee has reached the previous worker's level of skill, competence and speed (see chapter 11). All in all, losing valuable, experienced staff due to an unnecessary physical disability is a terrible waste that can be avoided in two steps:

1. Evaluating ergonomic risks
2. Designing workplaces that lessen the strain on the human body

Some potential causes of musculo-skeletal injuries are related to biological and lifestyle characteristics of individuals, and are therefore difficult to anticipate or do anything about using design. Biological and lifestyle-related factors influencing MSDs are shown in Table 2.1.

However, *work-related* MSD causes are possible for an engineer to avoid and are therefore the most interesting ones to identify quickly. Engineers with knowledge of ergonomics should design work and workplaces to minimize the adverse risks of the following:

- forced working postures
- load weight
- static work
- continuous loading of tissue structures
- repetitive working tasks
- time pressure/lack of recovery time
- working technique
- working attitude
- demotivation, stress
- organization

2.2. How big is the problem?

MSDs are the work-related health problem with the highest impact on sickness absenteeism in Europe; they are the cause of half of all absences from work and cost the EU €240 billion each year

Table 2.1: Individual biological and lifestyle-related factors that influence the risk of MSDs.

Biological factors	Lifestyle factors
• Muscular strength	• Prior load history, diseases and injuries
• Skeletal strength and bone mineral content	• Health, training and fitness habits
• Age, sex, biological measures	• Social environments (active or sedentary)
• Impaired vision, hearing, senses	• Pleasure, comfort and well-being
• Pain experience, neuromuscular reactions	• Chosen working postures
	• Smoking, alcohol, diet or drugs

in productivity losses (Fit for Work Europe 2013). MSDs are also the work-related health problem with the highest impact on permanent incapacity; 61% of permanent incapacity is due to MSDs (OSHA 2007). Forty-four million workers across the EU have an MSD caused by their work, 30% of those with MSDs also have depression, making it even more difficult for them to stay in or return to work (Bevan 2013). Typically the back tends to be the most commonly affected area of the body (Figure 2.1); 80% of all adults have back pain some time in their working life and 30% of sick

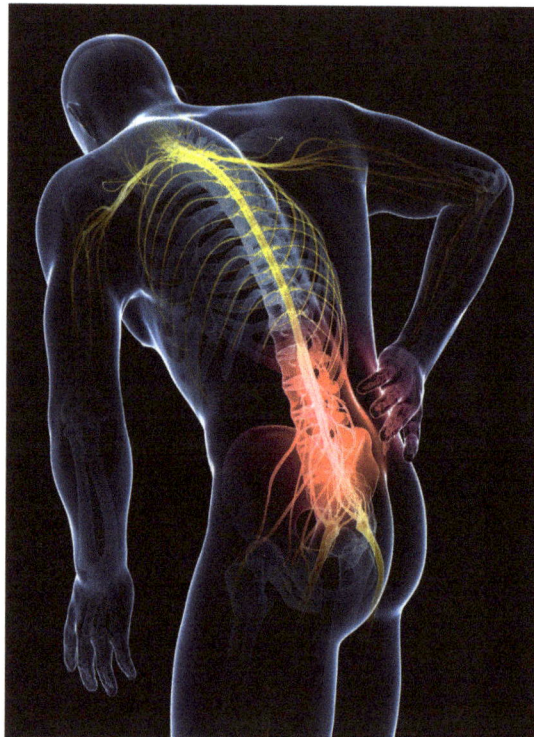

Figure 2.1: Lower-back pain has long been the most common cause of MSD-related sickness.

leave cases in Sweden are due to back pain (many young people) (Palmer, 2000). Blue-collar workers are at the highest risk for contracting an MSD, with almost 20 times as many employees experiencing an MSD compared to white-collar employees. Of these workers, those involved in manual labour such as trade workers, plant and machine operators and assemblers, are at the highest risk (OSHA 2007).

2.3. The musculo-skeletal system

The primary structures of the human *locomotive* (movement) system are the *skeleton*, the *muscles*, and the *joints*[2] (Figure 2.2). These structures combined allow the human body to move, withstand physical loading and recover when the body's abilities have been exhausted.

These are the structures that are mainly active when performing physical work, although other systems (such as the nervous system[3], the respiratory system[4] and the circulatory system[5]) that are all very important for the human being's ability to function are naturally also affected by physical work. However, this chapter will focus on movement, physical loading and what the locomotive system requires in order to function optimally.

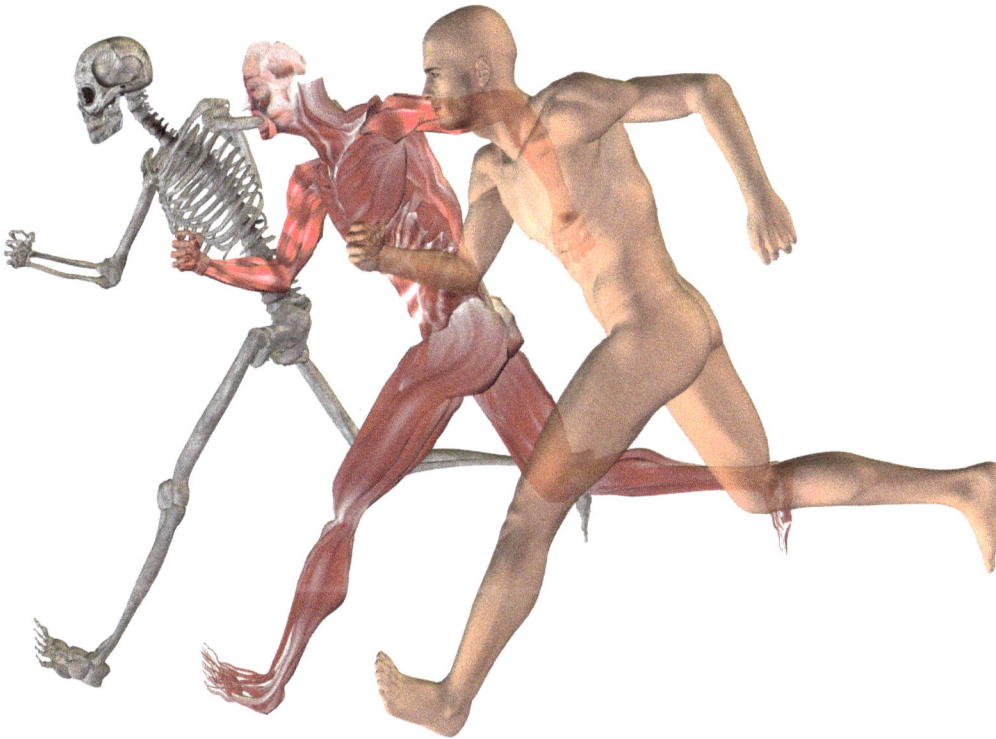

Figure 2.2: The musculo-skeletal (or locomotive) system consists of skeletal muscles, bones and joints.

Together, the skeleton, muscles and joints allow the human body to turn chemical energy into motion, to withstand physical forces and perform physical work in a way that is simultaneously dynamic, stable, flexible and adaptive. All of the structures are made up of living materials, so our body is constantly adapting to the loading and movements that we expose it to, making it better suited to perform those activities by becoming stronger and more stable. Unfortunately, it is also possible to load the body in such a way that we wear down or break the structures that make up our locomotive system. In order to avoid this and ensure that we design work and work systems that allow the human body to perform at its strongest, we need to know something about how each of these structures are shaped, how they move, how they respond to loading and regenerate.

2.4. The muscles

There are many different types of muscles in the human body, as shown in Figure 2.3. In the locomotive system, skeletal muscles convert chemical energy into contractions, thereby producing motion and mobility, stabilizing body positions, producing heat and helping to return deoxygenated blood to the heart. As the name suggests, most skeletal muscles are attached to the skeleton (via fibrous tissues at the ends called tendons) and are dedicated to moving it. This differentiates skeletal muscle tissue from cardiac muscle[6] and smooth muscle[7], which are not under our voluntary control.

Some skeletal muscles have specific functions, for example the postural muscles keep the posture of our body and head upright while we are awake, without any need for active control from our

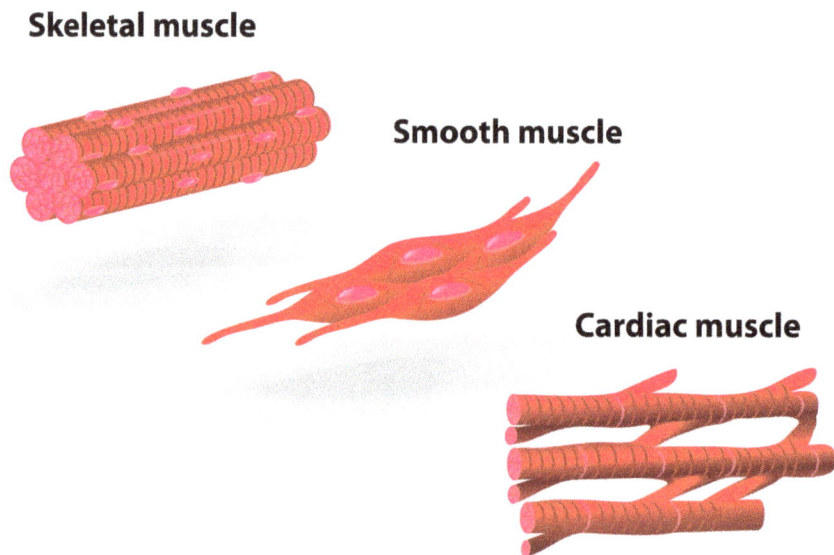

Figure 2.3: Muscle types: skeletal, smooth and cardiac.

brain (although in states of extreme fatigue, we lose control over our postural muscles, which explains the term "nodding off" to sleep). It is our skeletal muscles that allow us to transfer loads and torques, and the strength of our muscles varies depending on our age, sex, genetic heritage and training habits.

Definitions of how to count the number of muscles in the body vary, but they number in the hundreds (about 600 individual muscles) and they make up 40 to 50% of our body weight. Many muscles function in opposing pairs called antagonists, meaning that their contractions result in movements that work against each other. So when one antagonist is maximally contracted, the other one is – by definition – in a state of relaxation to allow the movement (see Figure 2.4). Examples of antagonists at work include bending and straightening of the knee or the arm, pointing and flexing the foot, or alternately bending the back outward and inward. For high-precision movements, the body controls a sophisticated and sensitive balance of contraction and relaxation between antagonists.

To stay balanced and well-aligned, the body generally needs to develop equal strength between antagonists; for example, some symptoms of back problems may actually have to do with weak stomach, or core musculature, rather than just the back muscles.

Healthy muscle tissue has four characteristics:

- *Excitability,* which is the ability to respond to stimuli
- *Contractility,* which is the ability to shorten and thicken (contract) when stimulated

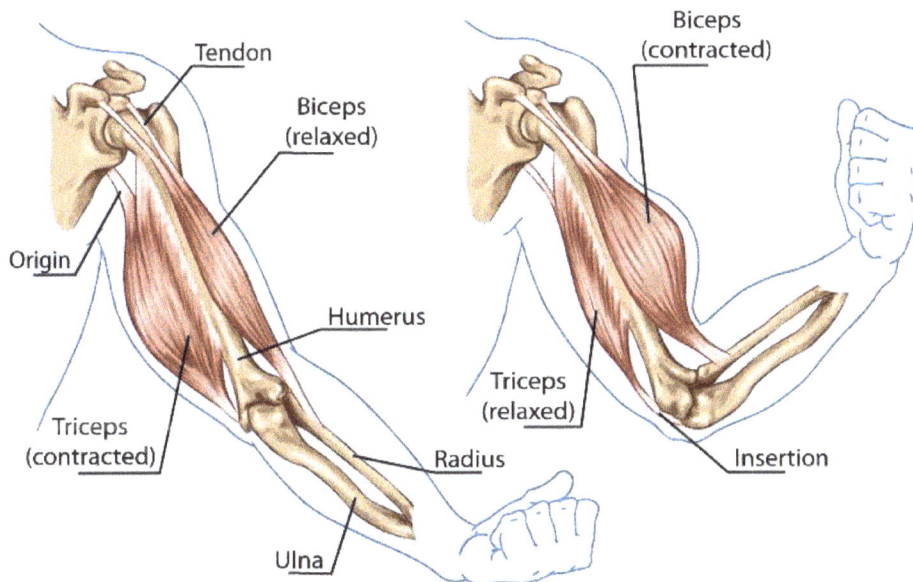

Figure 2.4: Antagonistic pair – biceps and triceps.

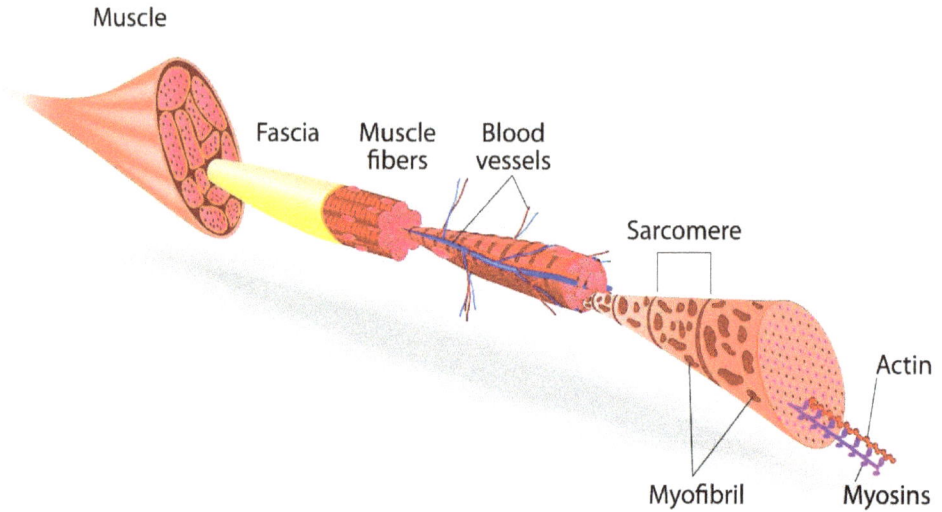

Figure 2.5: The structure of a skeletal muscle.

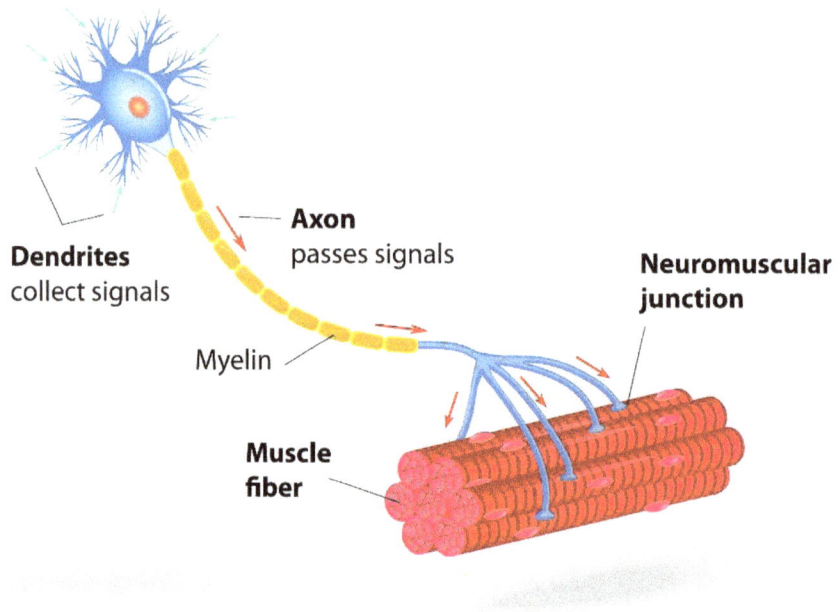

Figure 2.6: Structure of motor unit.

• *Extensibility*, which is the ability to stretch without being damaged
• *Elasticity,* which is the ability to return to its original shape after any form of physical loading

On a cellular level, muscles consist of clusters of long, thread-like cells called muscle fibres that measure about 2–150 mm long and 10–110 micrometres thick; see Figure 2.5.

Muscle fibres are in turn bundled into motor units, a group of cells that respond to voluntary signal impulses from the brain by contracting until the motor unit is fully, 100% contracted. This reaction of a motor unit (sometimes called "firing") can never be partial, so it is said that motor units are recruited one by one by the brain, until there are enough to perform the task. Motor units vary in how many muscle fibres they contain, and what type of movements and force generation they are adapted to. Each motor unit consists of one motor neuron and all the muscle fibres it contracts. The structure of a motor neuron is shown in Figure 2.6.

A *contraction* of a muscle can be explained as a chemical process that leads to a shortening and thickening of each motor unit, resulting in the production of force exertion and heat[8].

Generally, muscle fibres behave differently when stimulated by nerve impulses and can be classified into two types: Type I (slow-twitch fibres, suited for prolonged work and high endurance) or Type IIa or IIb (fast-twitch muscle fibres, suited for quick, explosive, brief movements). They are characterized by the type of physical loading or movement that they are best adapted to. Most people are born with a genetically determined proportion of Type I and Type II muscle fibres, but it is possible through physical training and nutrition to influence the proportions of different muscle fibre types. The differences in characteristics of these muscle fibres are described in Table 2.2.

Table 2.2: Main differences between different muscle fibre types.

Type I: Slow-twitch, aerobic	Type IIa: Fast-twitch A, intermediate	Type IIb: Fast-twitch B, anaerobic
• Adapted to high-endurance continuous contractions, work for a long period of time • Low force production • Slow contraction time • Small motor unit size • Not fatigued easily • Plenty of blood vessels and myoglobin = good supply of oxygen • Aerobic; requires plenty of oxygen to generate muscle fuel • Red in colour (due to high blood vessel content) • Predominant in marathon runners and cyclists	• Adapted to fast, short- term contractions • High force production • Fast contraction time • Large motor unit size • Quickly fatigued • Intermediate number of blood vessels and myoglobin • Mix of aerobic and anaerobic processes to generate muscle fuel • Red in colour	• Adapted to extremely fast, explosive contractions • Very high force production • Very fast contraction time • Large motor unit size • Fatigues very quickly • Low number of blood vessels and myoglobin • Completely anaerobic processes (no oxygen), burns glycogen to generate muscle fuel • White in colour • Predominant in sprinters, high jumpers

2.5. The skeletal system

The skeleton is made up of about 206 bones[9] (in an adult) which allow the human body to withstand its own weight with little or no muscular effort involved to stay upright and aligned (Figure 2.7). Apart from this, the most important functions of the skeletal system include:

- To serve as a rigid structure of mechanical stability, to support soft tissues and serve as attachment points for muscles
- To protect vital organs (brain, heart, lungs, spinal cord) and nerves
- To break down and regenerate bone (bone cells continually do this)
- To produce blood cells (in the red bone marrow)
- To assist in movement (skeletal muscles move bones) by making force and torque transfer efficient
- To store minerals (particularly Calcium (Ca) and Phosphorous (P))
- To store chemical energy (triglycerides, in the yellow bone marrow)

In a locomotive sense, the skeleton consists of a number of specialized bones suited for different purposes and loading profiles. The way the skeleton is designed, with an upright spinal column and long extremities with different bone widths and sizes, is the result of evolutionary requirements for human survival and development, in terms of structural strength, mobility, and flexibility. For example, the lower extremities (the legs) are quite wide and strong, and evidently suited for strength and stability in the lengthwise direction of the long bones (the femur over the knee, the fibula and tibia under the knee), greatly enabling us to stand, walk and run. Conversely, the arms (the upper extremities) consist of smaller, more complex bones that are developed to have maximum mobility and high precision, but (comparatively) low strength. This is because human survival has been highly dependent on our ability to move quickly and endure a lot of standing and walking, but also to use our hands as high-precision sensors and tools, causing a development of intricately attached small bones.

Some bones do not have the long shape and form of those in the extremities; some appear to be more like small, tightly clustered bones that are connected tightly and often form a base for complex-functioning body parts, particularly the bases of the feet and hands. In the hands, these bones are known as the *carpals* (*carpus* is Latin for wrist) and they form protective armour around a number of blood vessels, nerves and important tendons that allow finger movement. These all pass through a narrow passage in the wrist known as the *carpal tunnel*.[10] The corresponding clusters of bones at the base of our feet are called the *tarsals* (Latin for ankle). On both the hands and the feet, the bones that extend out to our fingers and toes are known as *phalanges*.

Since it contains blood, bone cells, energy and minerals, bone is a living material with a capacity to adapt itself to the type of loading it is under, and the body is continually breaking down or regenerating bone. It is essential to load the skeleton in order for bones to grow (this stimulates increased development of collagen fibres and more deposition of minerals, making the bones thicker and stronger) – if the bones are not placed under any type of stress, the body's processes of breaking down and reabsorbing bone materials overtakes the bone generation and the bones brittle and weak – a well-known phenomenon among old people, people who are bedridden long periods of time and astronauts due to weightlessness. This condition of bone fragility, where bone resorption processes outpace new bone development and mineral deposition, is known as osteoporosis. Such bones become so brittle and weak that a very small force application may break them, for example resulting in a hip fracture just from sitting down too quickly.

Frontal Bone

Orbit

Zygomatic Bone

Maxilla

Mandible

Cervical Spine

Clavicle

Acromion

Scapula

Coracoid Process

Humerus

Sternum

Ribs

Lumbar Spine

Radius

Ilium

Ulna

Sacrum

Pubis

Carpal Bones

Metacarpals

Phalanges

Ischium

Pubic Symphysis

Femur

Patella

Fibula

Tibia

Tarsal Bones

Metatarsals

Phalanges

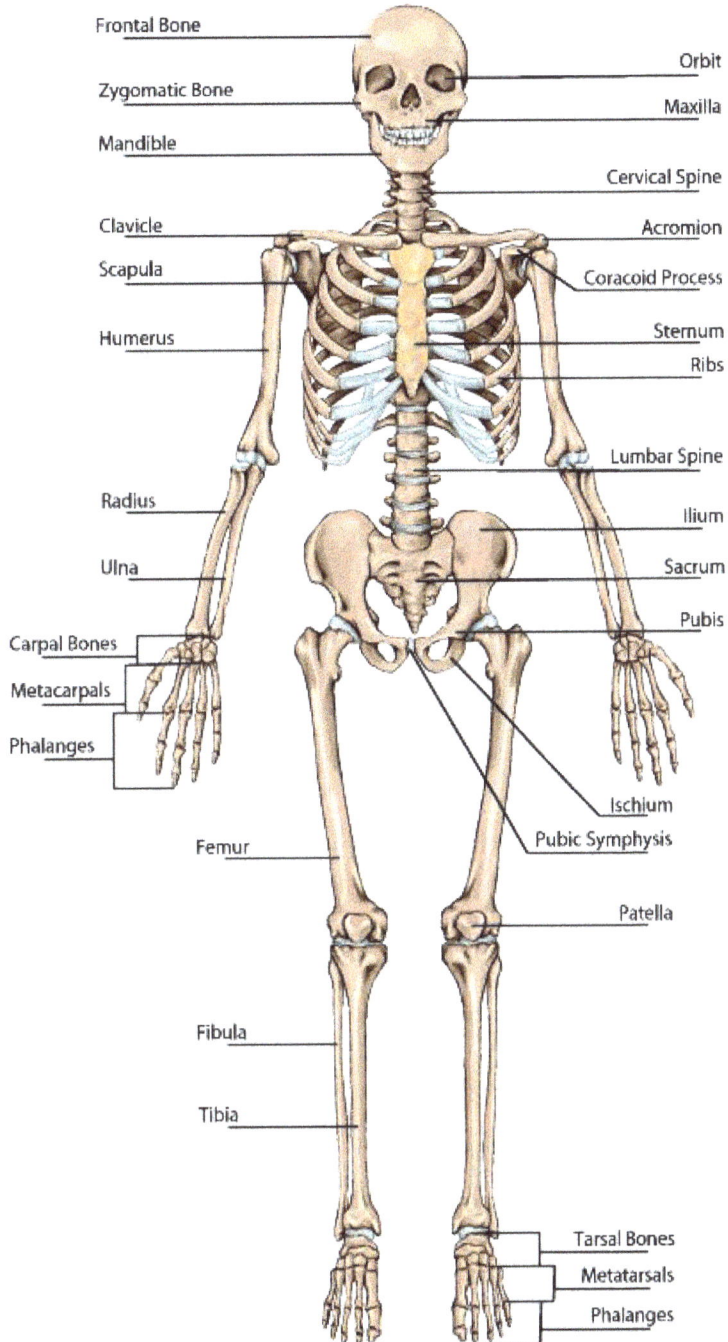

Figure 2.7: Structure of the skeleton.

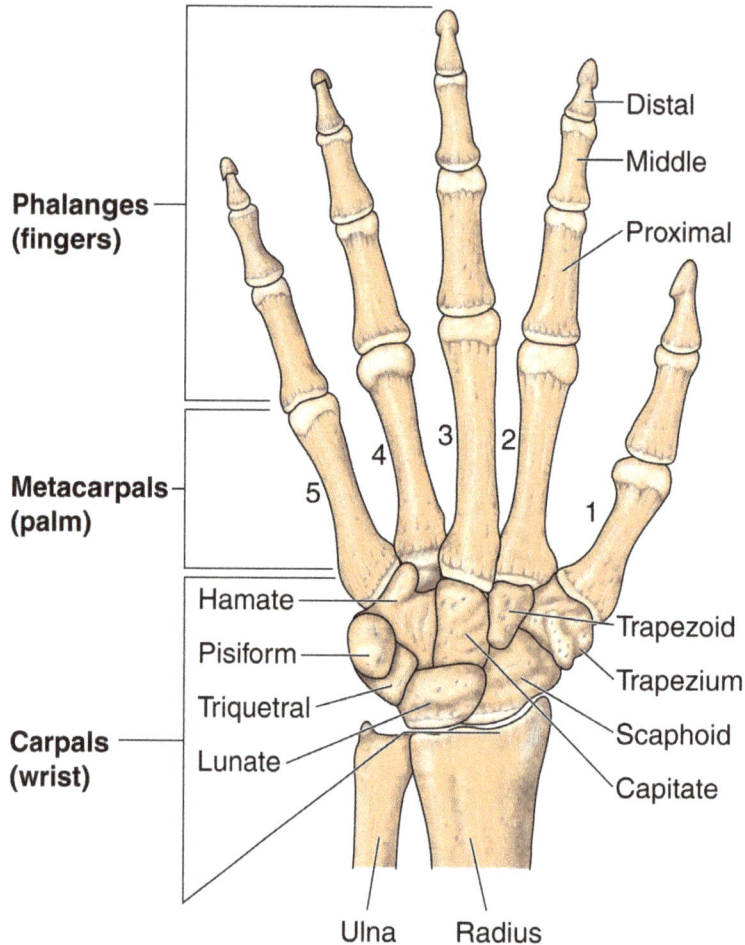

Figure 2.8: Anatomy of the hand.

2.6. Joints

Joints are the structures that appear at the points of contact linking bones to other bones, to cartilage or to teeth. Some joints are simply links between two bones without permitting movement at all, while others are specifically designed to permit movement, or at least a bit of flexibility[11]. Joints that allow movements in one dimension (translation, or "gliding" movement) may for example be found between the smaller bones in the wrist or where the ankle meets the foot. Two-dimensional joints, in many cases also known as hinge joints, allow rotation of bones relative to each other and are found, for example, in the elbows, knuckles and knees. Finally, three-dimensional joints permit the greatest range of movement in several dimensions, and are for example found at the base of the thumb (a so-called saddle joint) or at the shoulder and hip joints (ball-and-socket joints).

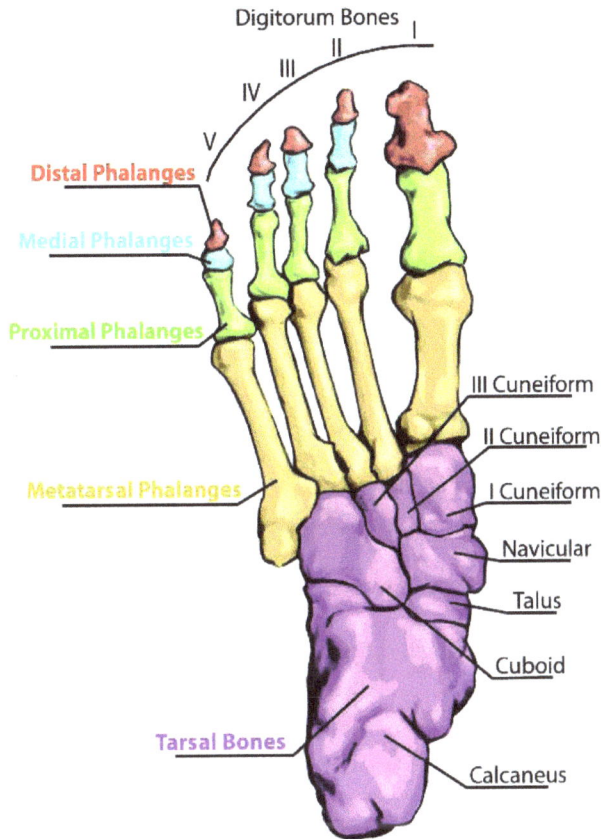

Figure 2.9: Anatomy of the foot.

The type of joint that permits movement in one, two or three degrees of freedom is called a synovial joint. Such joints are always between articulated bones (bones that meet and form a joint) whose ends are covered by a bendy, tough layer of articular cartilage, which reduces friction when the bone ends move (translate or rotate) relative to each other. Synovial joints always have a synovial cavity in which the bone ends move against each other, and are surrounded by a capsule filled with *joint fluid*[12] that lubricates the articular cartilage and allows even smoother movements with less friction between the bone ends. The capsule is covered by an outer layer of dense, tough connective tissue that is flexible enough to permit movement, but strong enough to keep the bones from dislocating. Depending on which joint it is, there may also be a presence of ligaments, which are bundles of fibrous connective tissue that are especially designed to withstand high strains.

Due to their complexity and the presence of many complicated and fragile structures passing through them, joints are particularly sensitive to injuries caused by physical loading in extreme positions.

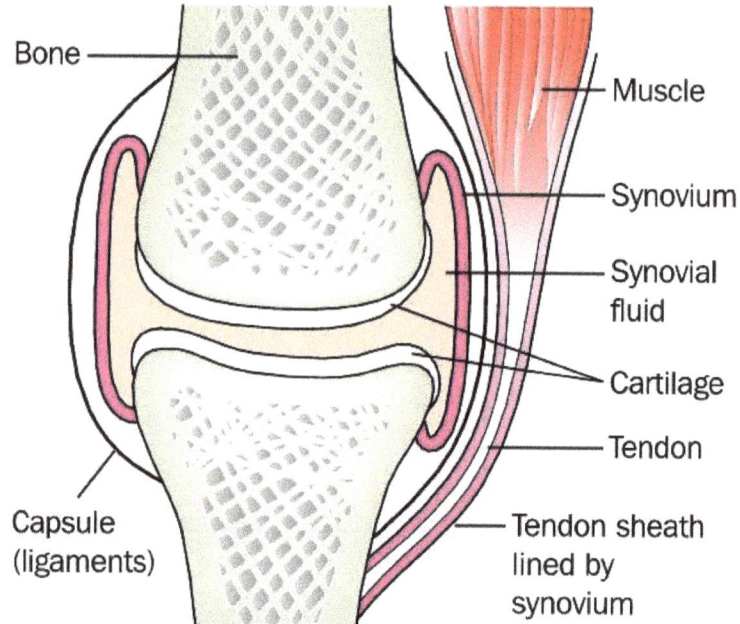

Figure 2.10: Structure of a two-dimensional joint.

Image reproduced with permission from Blamb/Shutterstock.com. All rights reserved.

The cartilage at the end of articulated bones is thickest in the middle, meaning that working in extreme joint angles may result in wear and tear on the thinnest part of the protective cartilage layer.
  As we age, the risk for joint problems and injuries increases due to a number of factors:

- Decrease in the production of synovial fluid, reducing both the lubrication of the joint cartilage and the transportation of unwanted substances away from the joint
- Articular cartilage between the bones becomes thinner
- Individual genetic and lifestyle factors
- Life-long wear and tear on the joint
- The fibrous ligaments around the joint capsules lose flexibility and become shorter, reducing the protection against movement-related injury and bone dislocation

2.7. Injuries and healing

When it comes to withstanding physical loading, the body is protectively structured in such a way that:

- The skeletal muscles protect the skeleton.
- The skeleton protects the inner organs.
- The joints protect blood vessels, tendons, muscles and nerves that run through them, but are also the most fragile structure of the three.

When any of these three structures are subjected to increasing mechanical forces, it is said that they are placed under *strain* until they can no longer withstand the force, and then they break. This stage is called *trauma*, and it means that the structures are injured and need time to heal before they can perform normally and take on more physical loading.

The muscle tissues are soft and can cushion the body (up to a point) from applied forces. The supply of plenty of blood flow, allowing the transport of nutrients and removal of unwanted materials, means the healing of mild to moderate muscle injuries typically takes place in a matter of weeks – more severe strains can take months.

The skeletal bones are excellently suited to withstand long-term loading in numerous directions (Figure 2.11) and static loads (such as the weight of our own body when we stand up), but because they have less blood flow and the required mineral deposits to create new bone take a long time to deposit, bones take longer to heal when injured. Injuries (breaks) in bone structures are called fractures, and usually heal in a matter of 5 to 6 weeks[13]. Additionally, healing times vary considerably depending on a number of factors, such as the injured person's age and general health, the site and severity of the fracture, the type of bone that has been fractured, proximity to a joint, infections, etc.

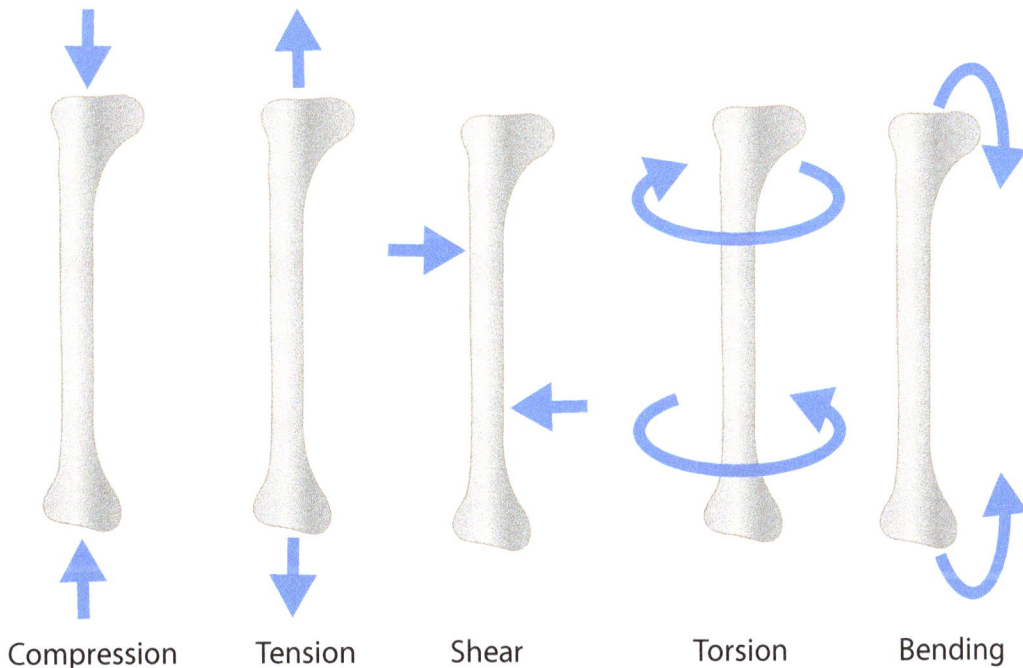

| Compression | Tension | Shear | Torsion | Bending |

Figure 2.11: Different types of mechanical loading on bones. The first two types act in the lengthwise direction of the bone, which it is well suited to withstand, but the others (shear, torsion and bending) bring greater injury risks.

Illustration by C.Berlin.

Joints are the most complicated and fragile structures of the three, partly because they consist of many different kinds of tissues and structures, but also because they are supplied with the least amount of blood flow (particularly ligaments). For this reason, injuries to joints can take months to years to heal, and depending on the age at which the injury is sustained, damages may be permanent. So, the priority order for work design is to avoid unnecessary risk of injuries first to joints, then the skeleton, then the muscles.

2.8. Movements

When the different functional tissues of the locomotive system work together, the body generates *movement*. The study of human movement is known as *kinesiology*. There is some useful standard terminology that is used in medical science to describe different types of movement, in terms of directions and orientation. Most human movements consist of bending or twisting motions that change the joint angles between different body segments. Some movements are coupled, in the sense that pairs of muscles work against each other to "do and undo" each other's respective movements (for example, bending and straightening the arm is, simply put, the work of two antagonistic muscles; the biceps and the triceps). The following terms are a helpful framework to describe motions.

Glossary For Movement, Organized by Coupled Motions[a]

CONTRACTION – when muscles use chemical energy and nerve impulses to pull together muscle fibre components so that the muscle becomes shorter and thicker, generating force exertion, movement and heat.

RELAXATION – when the contraction of muscle fibres is released, so that the muscle fibre components disengage from each other and the muscle becomes longer, more elastic and stops exerting force.

[a] Image permissions granted by: NoPainNoGain/Shutterstock.com (Contraction and Relaxation), stihii/Shutterstock.com (Antagonistic movement), and Kues/Shutterstock.com (Flexion). All rights reserved.

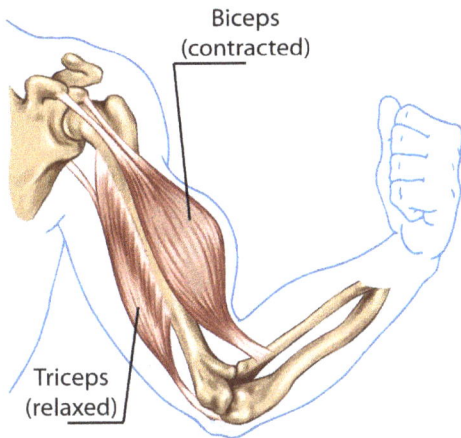

ANTAGONISTIC MOVEMENT – when a pair of muscles coordinate complex movement by working in opposite directions; when one contracts, the other releases.

FLEXION – a movement at a synovial joint that results in a decreased angle between two body segments (e.g. curling the biceps); alternately, think of it as curling in toward the body's midline.

EXTENSION[b] – a movement at a synovial joint that results in an increased angle between two body segments (e.g. straightening the arm); alternately, think of it as "uncurling", away from the body's midline.

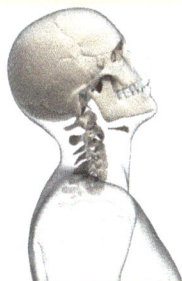

Hyperextension

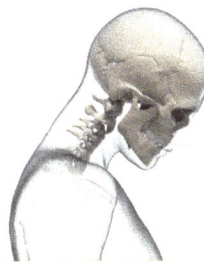

Hyperflexion

HYPERFLEXION and **HYPEREXTENSION** – flexion or extension "beyond anatomical position" towards the edge of movement range; usually beyond healthy loading limits (due to high pressure on the joints), but also usually prevented by resistance from the arrangement of ligaments and bones.

---

[b] Images reproduced with permission from: MetCreations/Shutterstock.com (Extension), Sebastian Kaulitzki/Shutterstock.com (Hyperextension and Hyperflexion), and C. Berlin (Abduction and Adduction).

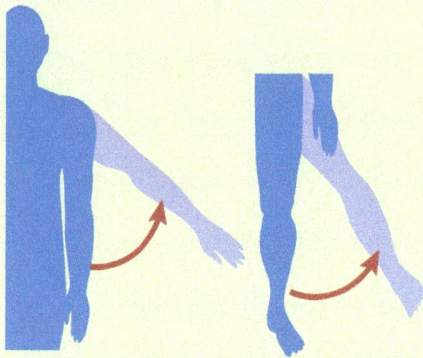

ABDUCTION – (for extremities, i.e. arms and legs) movement that brings the limb away from the body's midline, for example lifting the leg sideways.

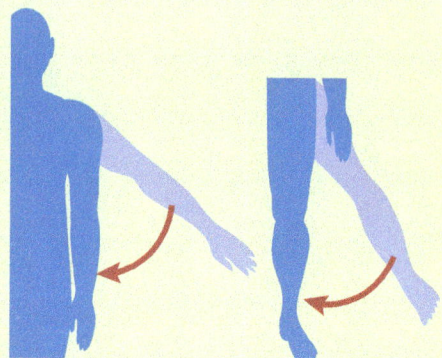

ADDUCTION – (for extremities, i.e. arms and legs) movement that brings the limb inward, closer to the body's midline.

(Right foot)[c]

Pronation
(leaning inward,
"flat foot")

Neutral

Supination
(leaning outward,
"rigid foot")

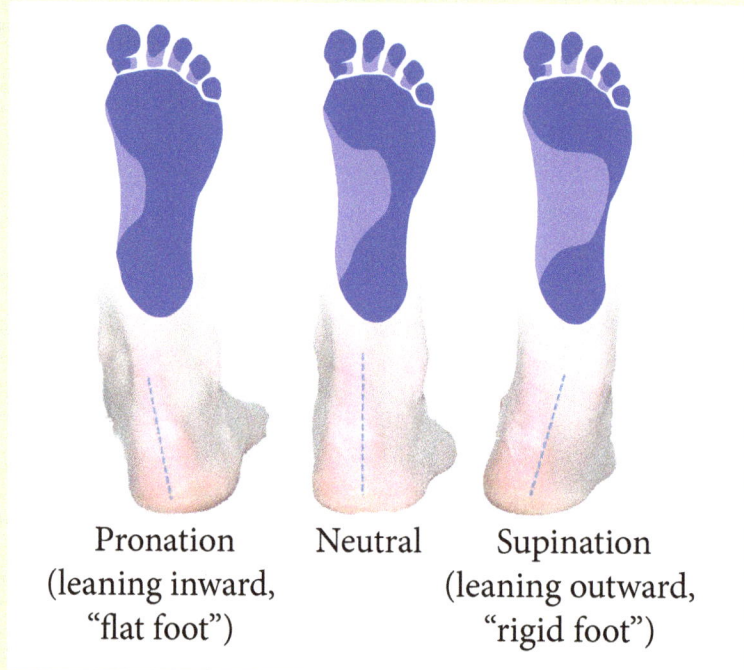

PRONATION – twisting motion
a) for feet: the ankle leans inward with weight pulling down the arch (also called "flat foot")
b) for hands: the palm is facing down and the radial (thumb) side is turned inward.

SUPINATION – twisting motion
a) for feet: the ankle leans outward, with weight on the outside of the foot and a high arch.
b) for hands: the palm is facing up and the ulnar (little finger) side is turned inward.

[c] Image by C. Adams and C. Berlin.

Another useful distinction between movements is whether they are *static* or *dynamic*. *Static* movement (or loading) usually means that the muscles' motor units are engaged for a long, sustained period of time (or in frequent repetition) without rest and recovery, until the point of fatigue (which occurs after a long time). Static movements are especially hazardous when they occur at low intensity, because then it is easy to ignore them or write them off as "not such a big load". Static work can involve keeping body parts still, small movements of part of the body for a long time while carrying out a task, or upholding an external load. Examples include working with arms above shoulder height, using a computer mouse and carrying heavy grocery bags.

Dynamic movement, on the other hand, is characterized by large, swiftly changing movements that may often involve great speed and/or large force exertions. While this type of movement may be a bigger risk for sudden trauma to the locomotive tissues (such as torn muscles or ligaments), this type of loading is also characterized by much more loading variation, leading to relatively frequent rest and recovery while different muscles take turns being loaded. In comparison, static loading can gradually wear down locomotive abilities due to the constant loading of the same muscles, pressure on the same body structures, etc.

A good rule of thumb from a health perspective is that both these movement types can help the body become stronger and more prepared for high loading, provided that there is sufficient rest and recovery for the body to replenish the needed oxygen and energy via the blood, and to transport waste products from the locomotive tissues. This is what distinguishes risky workloads from intentional physical training: although both can push the body to its loading limits, a work environment may sometimes allow little or no time for recovery, while physical training is designed to alternate between loading the body and letting it recover.

Simply put: Dynamic loading with breaks and variation is mostly good and strengthening, while static loading with few breaks risks being harmful and weakening.

2.9. Musculo-skeletal complexes

As previously mentioned, certain parts of the body are frequently represented in MSD injury statistics. It is worthwhile to get to know some of these musculo-skeletal complexes and their strengths and weaknesses better.

2.10. The back

The back is a complex entity consisting of active and passive tissues. The active tissues (skeletal muscles) voluntarily move the back by bending or rotating it, and the passive tissues (bones, joints, ligaments and intervertebral discs) take up structural and external loading.

The spine

The spine is made up of a series of stacked bone structures called vertebrae (singular: vertebra). Between each vertebra is a layer of tough fibro-cartilage that encases a soft, gelatinous, highly elastic disc (called an *intervertebral disc*). These discs allow extra movement flexibility and help the body absorb vertical shocks. Together they form a flexible, strong column, whose moveable parts (Figure 2.12) are usually grouped into three sections: the cervical (neck) spine, the thoracic (chest cavity) spine and the lumbar (lower back) spine.

The spine's function is to:

- Protect the spinal cord and nerves (which run through it from the brain to the rest of the body).
- Support and hold the head.
- Allow trunk mobility by being able to rotate and bend forward, backward and sideways.
- Transfer loads and torques from things we push, pull, carry and lift.
- Serve as a point of attachment for the back musculature and ribs.

Whenever we are awake and active, the weight of our body and our activities cause *spinal loading*, which leads to compression of the intervertebral discs. As long as the body is erect and the spine is in its natural S-curve shape, the discs can take this very well since they are loaded with even and symmetrical pressure. However, if the back is bent and subjected to external loading (such as when

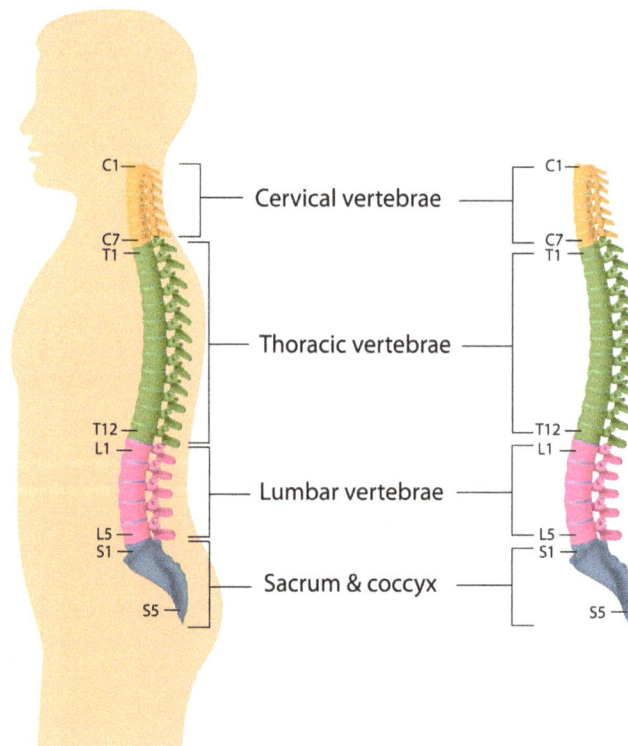

Figure 2.12: The cervical, thoracic and lumbar regions of the spinal column. The sacrum and coccyx – also known as the tailbone – are not part of the moveable spine.

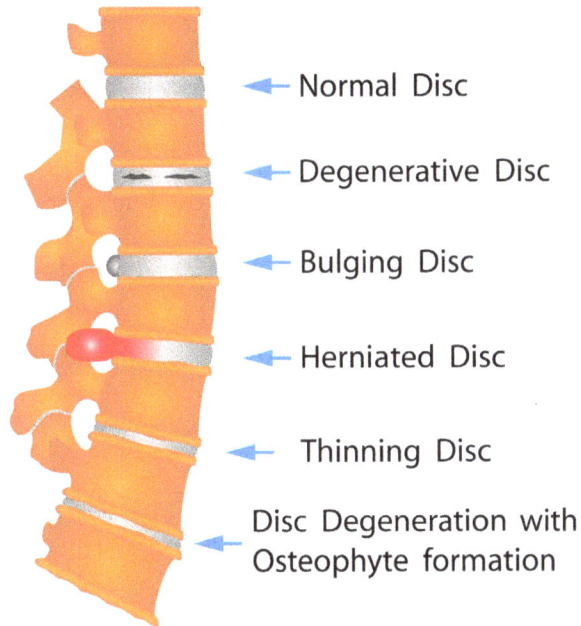

← Normal Disc

← Degenerative Disc

← Bulging Disc

← Herniated Disc

← Thinning Disc

← Disc Degeneration with Osteophyte formation

Figure 2.13: Different spinal disc conditions.

a person's back is bent and they are trying to lift something up), the discs are compressed with high pressure only on one edge.

In a worst-case scenario with too-high force, wear and tear from daily loading and/or age, the fibro-cartilage casing can bulge or break and the gelatinous disc can rupture, a condition sometimes referred to as a *herniated disc*[14] (Figure 2.13). This can cause great pain and/or numbness if the disc rupture comes in contact with a nerve root or the spinal cord; but the condition normally settles by itself (by the body reabsorbing the disc fragments) in a matter of one to six months.

Sitting, standing and lying down

The natural shape of the spinal column, when we are standing, is an S-shaped curve when viewed from the side. This shape is possible when there is no imbalance, twisting or bending in any direction; it occurs when the head, hips and feet are vertically aligned and symmetrical. When the spine is positioned this way, the passive structures (the vertebrae, ligaments and discs) are at their strongest alignment and the body is in its absolute strongest condition (from a posture perspective) to take on physical loading. Posture strongly influences the spinal loading and disc compression, in terms of loading on the lumbar (lower) spine, the difference between standing and sitting is significant. According to Kroemer and Grandjean (1997 p. 73), if we normalize the loading from standing in an erect, relaxed position with the naturally occurring S-curve to 100% compressive loading, then relative to that pressure in the lower back:

- Sitting down with a straight back corresponds to 140% loading.
- Sitting down in a slouch or leaning forward corresponds to up to 190% loading.
- On the other hand, lying down brings down the compressive loading to 24%.

This difference in compressive loading provides a clue to why it is so important to get sufficient amounts of sleep – not only does the body need to recuperate and the mind work subconsciously with processing information, the discs in our back also need a chance to return to their uncompressed, round form in order to be ready for another day's work as flexibility enablers, loading relief and shock absorbers.

2.11. The neck

Technically, the neck is a very flexible continuation of the cervical (upper) spine (Figure 2.14). It connects our head to the shoulder complex, allows flexibility of movement, and is an attachment point for several small and large postural muscles that keep our head erect. However, the neck deserves special attention since it is an area of the body that is highly complex, sensitive and prone to injury. An injury to the neck may cause severe impairment, since many nerves run through it. The head weighs about 8% of a human body's weight (about 4.5–5 kg in an adult), and is a special condition of loading for the neck, especially in cases where the head is not held erect.

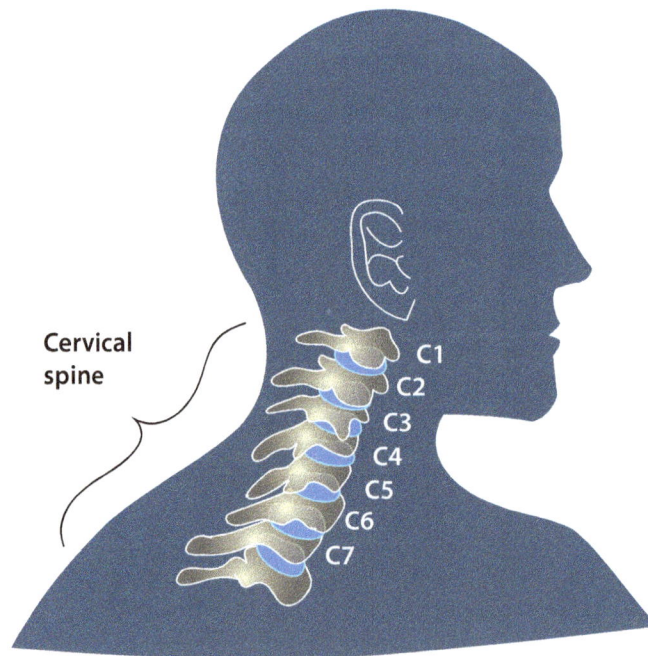

Figure 2.14: The cervical spine (C1–C7) that supports the neck.

Illustration by C. Berlin.

Neck injuries and problems tend to arise due to the following situations:

- Frequent or static bending for long periods of time – for example when using smart phones or tablets – causing shear forces between the vertebrae, as well as high muscular tensile (stretching) loads.
- Extreme extension of the head causing biomechanically dangerous loadings.
- Whip-lash injuries (often caused in traffic accidents) – hyperextension followed by sudden hyperflexion.

The neck and shoulders, being so closely connected, are often co-dependent in their function and movements, and work-related tension or pain in either structure may easily spread to or affect the other. Small injuries that propagate this way often lead to compensation by the other structures, resulting in even greater pain. Quite frequently, a combination of neck and shoulder problems is overrepresented in WMSD statistics, and long periods of sick leave are required. Some occupational groups at risk for neck-shoulder complex injuries are often exposed to static and repetitive loads when working; such as computer and office workers, cashiers, light manufacturing industry workers, health service workers and truck drivers.

2.12. The shoulders

The shoulder is a very complex anatomical structure that allows great freedom of movement, but is also sensitive to developing pain and injuries. Since it is connected to large muscles both in the front and back of the body, as well as several more weak muscles connected to the neck and arms, the bones of the neck-shoulder complex are completely dependent on the balance and alignment of the muscles and fascia (binding tissue) that tie them together.

The arm/shoulder joint is flexible in three dimensions thanks to the presence of four different joints in the shoulder area (Figure 2.15).

- The joint at the arm-to-shoulder connection, a ball-and-socket joint called the *glenohumeral* joint.
- The joint between the collarbone (clavicle) and the top of the shoulder blade (scapula), called the *acromioclavicular* joint.
- The joint between the collarbone and the middle of the ribcage (the sternum), called the *sternoclavicular* joint.
- The joint between the shoulder blade (scapula) and the back *scapulothoracic* joint (where the scapula meets with the ribs at the back of the chest).

The arm joins the upper body in the shoulder area, where it can be abducted, adducted, rotated, flexed and extended. The movement ability is dependent on the healthy function of the joints, muscles, neck and spine. In particular, the ball-and-socket joint between the shoulder and arm is the type of joint that allows movement with 3 degrees of freedom, but this comes at a cost; since the head of the arm bone only has a shallow fit in its socket, meaning that the main stability in the shoulder depends on the rotator cuff muscles and shoulder ligaments to stay stable. Since joints are the most injury-sensitive tissue in the locomotive system, this unfortunately also makes the shoulder as a whole vulnerable in terms of being easily dislocated, inflamed or worn out. It is stabilized by the collaboration of weak muscles, ligaments and the ball-and-socket joint.

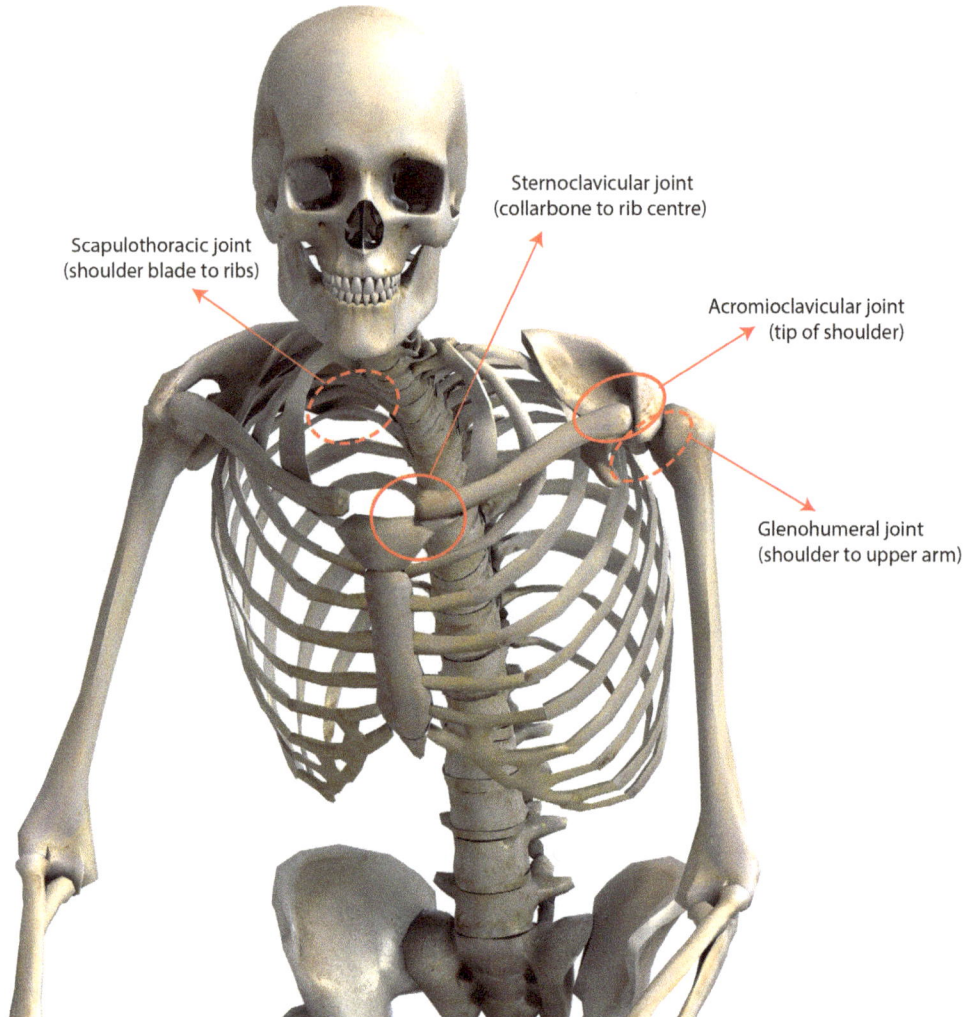

Figure 2.15: Structure of shoulder joints.

Illustration by Ralf Juergen Kraft with modifications by C. Berlin. Permission granted via Ralf Juergen Kraft/ Shutterstock.com. All rights reserved.

If pain or injury starts in one part of the shoulder, it is likely that surrounding structures start to compensate by tensing up. This leads to static loading, which in turn can result in pain and discomfort spreading to other parts of the body than where the pain originated. At worst this can result in inflamed, swollen tissues, decreased blood flow, decreased freedom of movement, deformation of muscles and tendons, and impingement (constriction or squeezing in tight spaces) of nerves. Several small muscles and tendons also run through narrow spaces in the shoulder joint and can be vulnerable to high pressure caused by posture.

Figure 2.16: There are many ergonomic pitfalls that may cause injury in the workplace.

Some classical pitfalls that may cause shoulder pain and injuries include the following (Figure 2.16):

• forward flexion of the shoulders
• work with arms above shoulder level
• work with arms outside the body area
• raised shoulders
• repetitive work
• static work loads
• prolonged work with low static loads

2.13. The hands

The hands and wrists are particularly critical for a human being to be able to work. For most of us, the hands are a tool for working, sensing and self-expression. The hand, wrist and arm form a complex and sensitive structure together, that can get easily overloaded or injured during physical work. An injury to the hands has serious consequences since it generally hinders human beings from carrying out most types of work.

The skin of the hand has 17,000 receptors for sense (including cold and heat receptors and nerve endings, some of which are stimulated by fine hairs), which allow us to respond to touch, pressure, pain, heat and cold by adapting our exerted force and movement precision. The hands are also used to convey emotions, personality and body language; imagine (or better yet, try) a conversation where the hands are not used. In most social settings, this would remove an important dimension from communication and probably be considered rather odd behaviour.

The bones, muscles and joints of the hand are primarily adapted for high-precision work and are not anatomically suited for exerting high force[15]. Therefore, it is very important to design work that allows the hand the best possible conditions to exert force and precision. This includes the correct design of hand tools, for comfort, skill and precision during work. The possible motions of the hand (see Figure 2.17) include flexion and extension (for both the fingers and the wrist), deviation (sideways wrist bending) and the twisting motions *pronation* and *supination*.

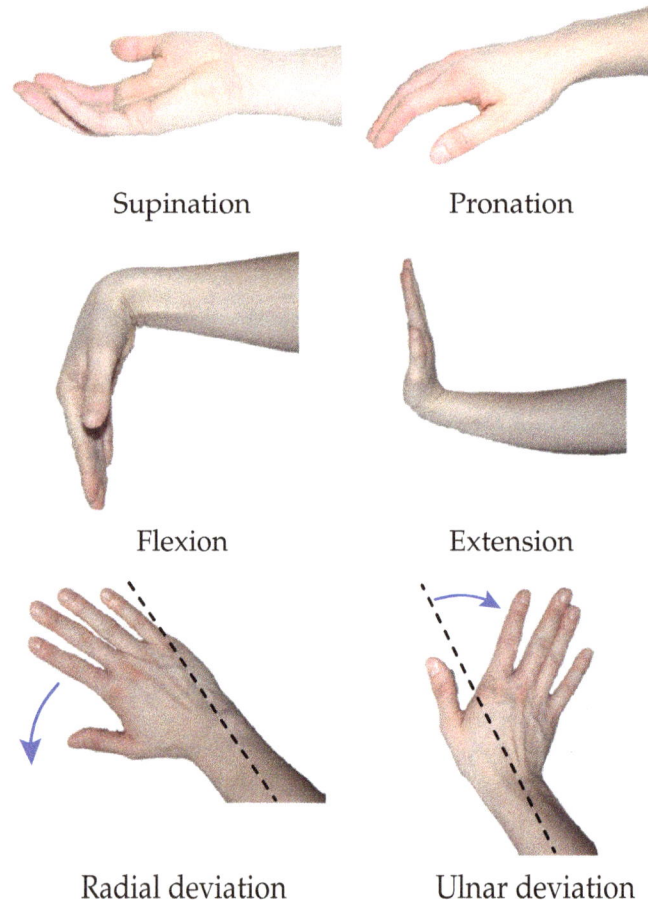

Figure 2.17: Motions of the hand and wrist.

Images by C. Adams. All rights reserved.

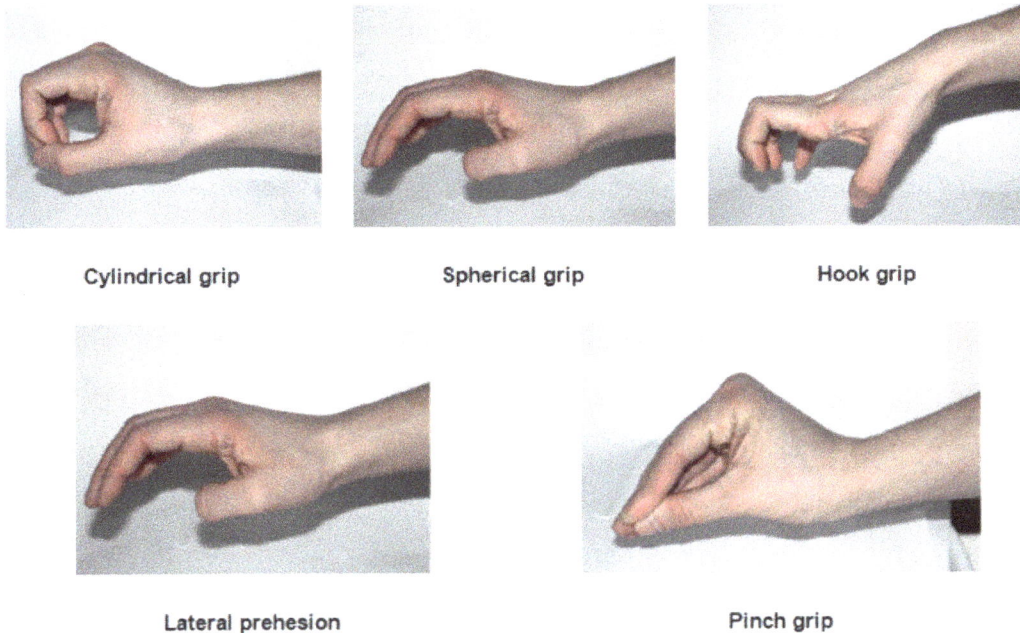

Cylindrical grip   Spherical grip   Hook grip

Lateral prehesion   Pinch grip

Figure 2.18: Overview of the gripping functionality of the hand.

The hand is also an important grip tool that has different functional positions depending on what level of power and precision is needed for a task. See Figure 2.18 for an overview of grip types.

It is important not to overload the complex structures of the hand by unnecessary twisting and bending[16] while working or handling loads. The hand has a functional resting position, in which the wrist is straight, the muscles are relaxed, the fingers lightly curled, and the pressure in the carpal tunnel (the narrow passage in the wrist that encases the median nerve and several tendons) is at its lowest. As much as possible, work for the hands should be designed as close to the functional resting position as possible, since both the strength and the precision of the hand decrease drastically at the extreme ends of our movement range, as shown in Figure 2.19 and Figure 2.20.

Some typical work-related problems that may result in injury or impairment of the hand's function include:

- repetitive tasks
- high forces
- punctual pressure on a small area
- incorrect grips
- vibrations
- cold and heat
- extreme positions during work (e.g. ulnar deviation combined with supination)
- incorrect design of hand tools

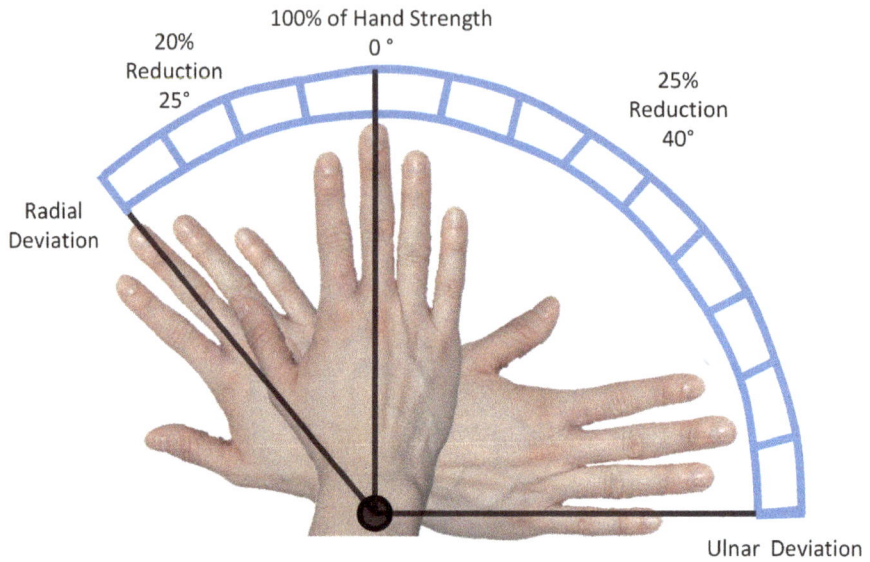

Figure 2.19: Reduction of hand strength at different angles of deviation.

Images by C. Adams. All rights reserved.

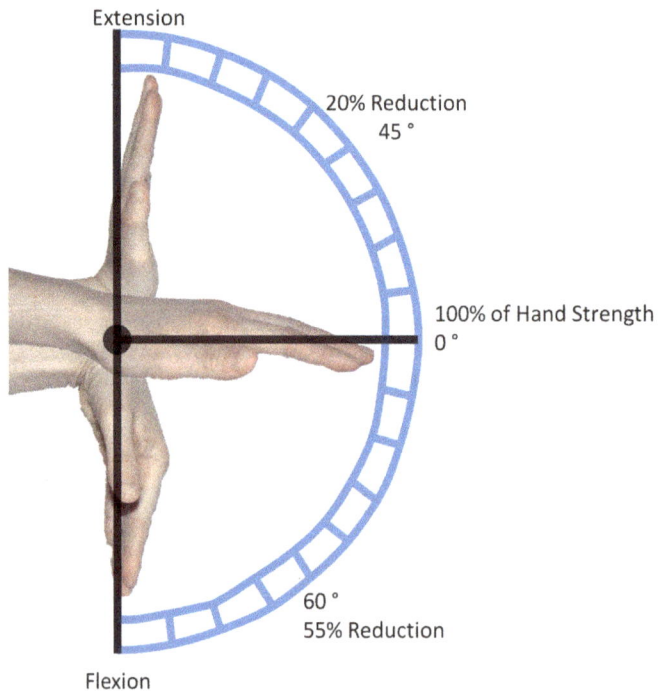

Figure 2.20: Reduction of hand strength at different angles of flexion/extension.

Designing hand tools

When working with hand tools, the hand should be as close as possible to its functional resting position, to ensure good conditions for strength and precision development. Good design that focuses on this minimizes the risk for long-term consequences like injuries and discomfort. It is also worthwhile to consider the context and working environment for hand tools; for example, the ability to grip a tool is affected by the use of gloves or protective clothing; by temperatures that make the tool uncomfortable to use; by vibrations; or by substances or humidity that might make surfaces, materials and tools wet, slippery and/or dirty. Reduced friction may significantly reduce grip and control over a tool. For example, medical equipment should be able to withstand exposure to blood and chemicals, while wooden handles on tools may be excellent for bare hands, they aren't for gloved ones.

When designing hand tools, ask these questions:

- Who is going to use the tool, and for what purpose?
- What is the function – what task is to be solved?
- Are there differences to consider in the design population, e.g. between sexes (male/female grip strength ranges) or cultures (preferred hand for different activities)?
- What anthropometric data is useful (e.g. different sizes, left/right hand prevalence etc.)?

HAND INJURIES	
Carpal Tunnel Syndrome (CTS)[a] This condition is caused by highly repetitive work in extreme positions, in combination with high force development. The median nerve, which runs through the space in the wrist called the carpal tunnel, gets pinched due to increased pressure. This leads to numbness, tingling, decreased function and weakness in the area around the nerve, and the fingers that are affected by the median nerve (the thumb and three middle fingers). Similar symptoms can be had for radial and ulnar nerves. Treatment of CTS depends on the severity, but non- surgical treatment usually includes wearing a supportive wrist splint to prevent the wrist from bending.	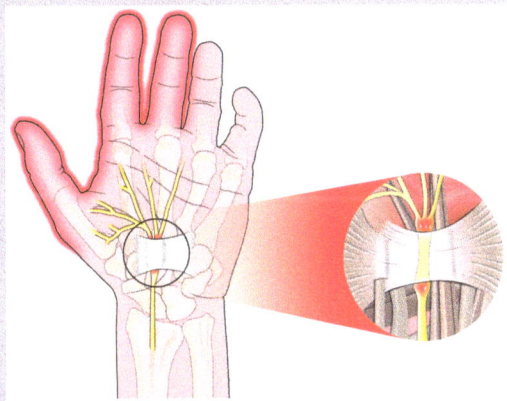
Inflammation in tendons[b] Tendon inflammation is a condition where movement of the wrist and fingers is painful due to a sense of pressure and swelling at the knuckles. It is the result of irritation in the tendons' sheaths, caused either by highly repetitive finger work or sharp edges on hand tools. One symptom, known as "trigger finger syndrome", is an inability to flex and extend the thumb and forefinger in one smooth movement – instead, the movement is hindered until it "snaps" into position. Medical language distinguishes between *tendonitis* (inflamed tendon) and *tenosynovitis* (inflamed tendon sheath).	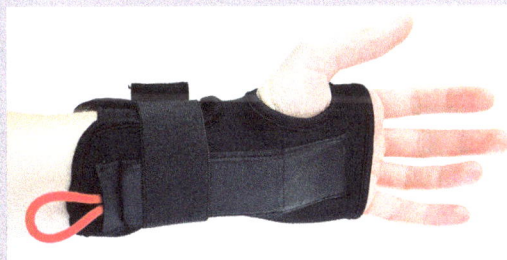

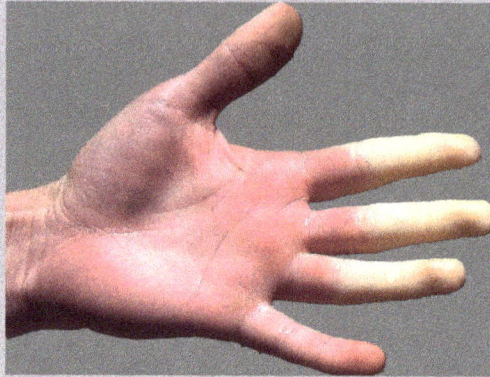

White fingers[c]

"White fingers" is a condition with numb, tingling fingers, where blood flow is so decreased that the fingertips turn white. The condition may be hereditary, in which case it is called *"Raynaud's Syndrome"* (therefore, it is important to determine the individual's medical history) or it may be the result of a MSD caused by hand-arm vibration. The greatest risk for contracting this injury occurs at frequencies between 50 and 150 Hz. When caused by work, it is also called *vibration white finger* (VWF), *hand-arm vibration syndrome* (HAVS) or *dead finger*. However, it is sometimes hard to distinguish from Raynaud's syndrome, where the characteristic white fingers appear due to biological or non-work- related causes. Symptoms include: discoloured, pale white fingertips, especially in cold temperatures; numbness and prickling sensations in the fingers; and a decreased motor function and sense of touch.

Study questions

Warm-up

Q2.1)   What are the three main tissue structures of the human locomotive system and what are their functions?

Q2.2)   What structures make up the active and passive parts of the spine?

Q2.3)   Why is it important for the back to get a good night's sleep?

Q2.4)   Why is the shoulder area particularly sensitive to joint injury?

Q2.5)   Why should you try to lift heavy objects with your leg muscles rather than your back?

Q2.6)   Why is it risky to perform physical work in extreme positions?

Q2.7)   Why is it not considered ergonomically risky to perform strenuous physical exercise with maximal force exertions and working until fatigued, as you would at a gym?

Look around you

Q2.8) Observe someone performing physical work (for example in a cash register, in a shop, at a gym etc.) – look particularly at the back, the neck, the shoulders and the hands. Is the person performing the work with a good posture, or can you see any signs of asymmetrical loading?

Q2.9) Clench and open your fists and wiggle your fingers – where is the majority of muscle activity happening as you do this? Try feeling the muscles in the palm of your hand and the underarm while you activate your hands. Where do you think you will feel fatigued if you exert large forces with your hands?

Q2.10) Hold a pen in the palm of your hand and grasp it in your fist. Then try holding it in extreme flexion and extreme extension – what happens to your ability to grasp the pen tightly?

Connect this knowledge to an improvement project

- When you observe physical work for the first time, try to take note of what movement types (bending, twisting, pushing, pulling, lifting, precision movements) and strength levels (in the back, arms, hands) are required to perform the work to good quality.
- Try to assess if body structures are loaded properly – as in symmetrically, at appropriate force levels and not to the point where they get fatigued.
- Do you observe any risk for fatigue or force overloading?
- Reflect on if physical work demands are appropriate for all ages, sizes and physical conditions of your working population. Who should be able to perform this task? Identify any "critical users" who may not be able to do the task currently.
- Look particularly at hand loading and tools – are they appropriate for all workers? Can anything in the tools be improved to lessen the demand for extreme postures, large force exertion or long exposure times?

Connection to other topics in this book

- All of the theory in this chapter is the foundation for the chapters on Physical Loading (Chapter 3), Anthropometry (Chapter 4), Ergonomics Evaluation Methods (Chapter 8), and Digital Human Modeling (Chapter 9). All of the principles in those chapters are rooted in the basic rules of how much loading the anatomical structures can withstand.

Summary

- The human body is a very complex structure made up of bones, muscles and joints; if loaded in the wrong way it can easily get injured.
- Combined, the skeleton, muscles and joints enable the body to turn chemical energy into motion, withstand forces and perform physical work.
- With the knowledge of basic physical anatomy and how certain structures move and respond to loading, it is possible for engineers to design healthy workplaces with reduced risk for injury.
- Work-related injuries resulting from repetitive static tasks and heavy loading are unfortunately quite a common occurrence, with the highest impact on employees taking sick leave in Europe.
- To avoid pain, discomfort, fatigue or injury the body should be used in its natural position, as close to neutral as possible.
- Most skeletal muscles are attached to the skeleton and enable humans to transfer loads and torques, while protecting the skeleton. Their strength is dependent on age, gender, genetic heritage and training.
- There are two types of muscle fibres: fast twitch and slow twitch. Fast twitch are suited to short fast explosive contractions while slow twitch is better for sustained longer exertions.
- An adult skeleton is made up of 206 bones of varying size and function.
- Joints are structures positioned at the point where different bones connect; they can enable movement in up to three different dimensions.
- Joints are the most complex of the three structures and can take years to heal if injured, or in some cases never fully heal.
- The back is one of the most common areas affected by WMSDs. The spine is made up of a series of stacked vertebrae and discs.
- When sitting or standing the back is being loaded and the discs between vertebrae compress. Excessive or uneven loading can cause discs to rupture, resulting in severe pain or numbness.
- The neck and shoulder complex are also a common area affected by WMSDs. Frequent or static bending of the neck resulting from looking at screens is a common injury trigger.
- The hands and wrists are crucial for carrying out high-precision work tasks, and an injury here has serious implications as it hinders humans from most forms of work.
- The hand and wrist can move in a number of different directions; however, working with them as close to the functional resting position as possible enables the best performance conditions for high strength and good precision.

Notes

[1] According to Kuorinka and Forcier (1995), the term *work-related musculoskeletal disorder* excludes accident-related sudden injuries.

[2] If you want to learn more about anatomy and physiology in a more medical sense, the book *Introduction to the Human Body* by Tortora and Grabowski (2004) is warmly recommended. Please see the references at the end of the chapter.

[3] The brain and nerves

[4] The lungs and oxygenation of the blood

[5] The heart and blood flow

[6] Pumps blood to and from the heart

[7] Transports food and liquid through the gastrointestinal (digestive) system

[8] For this reason, when the body needs to increase its temperature, we shiver involuntarily.

[9] This number may vary, partly due to age, partly due to different conventions of how to count bones in the skull, and partly because some individuals are born with superfluous bones, e.g. extra ribs or vertebrae.

[10] See the fact box in section 4.3.4 on hands to read about the condition *carpal tunnel syndrome.*

[11] There are actually six defined types of joint movements defined by the anatomical structure of the joint, but in this book we simplify it to the principle of movement in one, two, or three dimensions.

[12] Also called synovial fluid; it is secreted by an inner synovial membrane in the joint capsule.

[13] However, this is just the structural recovery of the bone; the healing time until the bone is ready to take on the same amount of loading usually requires an extra period of rehabilitation.

[14] Also referred to sloppily as "slipped disc", although this condition does not actually mean that the disc slips per se; it is still a rupture of the gelatinous core.

[15] Although the hand (like any other body part) can be deliberately trained to exert high forces given the right exertion-and-relaxation regimen, it is unsustainable to require very high grip strength of a working population that you are designing for.

[16] For as you now know, that would mean working at the outer extremes of your joint motion range, where the joint cartilage is thinnest and the internal pressure is highest.

2.14. References

Bevan, S. (2013). Reducing Temporary Work Absence Through Early Intervention: The case of MSDs in the EU. [Online]. Available from: http://www.theworkfoundation.com/DownloadPublication/Report/341_The%20case%20for%20early%20interventions%20on%20MSDs.pdf [Accessed 4th December 2013].

Fit for Work Europe. (2013). MSDs. [Online]. Available from: http://www.fitforworkeurope.eu/Images/MSD_infographic_screen.pdf [Accessed 4 December 2013].

Kroemer, K. H. E. & Grandjean, E. (1997) *Fitting the Task to the Human: A Textbook of Occupational Ergonomics.* London; Bristol, PA: Taylor & Francis.

Kuorinka & Forcier (Eds.). (1995). *Work Related Musculoskeletal Disorders (WMSDs): A Reference Book for Prevention.* London: Taylor and Francis.

OSHA. (2007). MSDs. [Online]. Available from: http://www.osha.mddsz.gov.si/resources/files/pdf/E-fact_09_-_Work-related_musculoskeletal_disorders_-MSOs-_an_introduction.pdf [Accessed 4 December 2013].

Palmer, K. T., Walsh, K., et al. (2000). Back pain in Britain: comparison of two prevalence surveys at an interval of 10 years. *BMJ*, 320: 1577–1578.

Bibliography

Tortora, G. J. & Grabowski, S. R. (2004). *Introduction to the Human Body: The Essentials of Anatomy and Physiology,* 6th edition. Hoboken, NJ: J. Wiley & Sons.

CHAPTER 3

Physical Loading

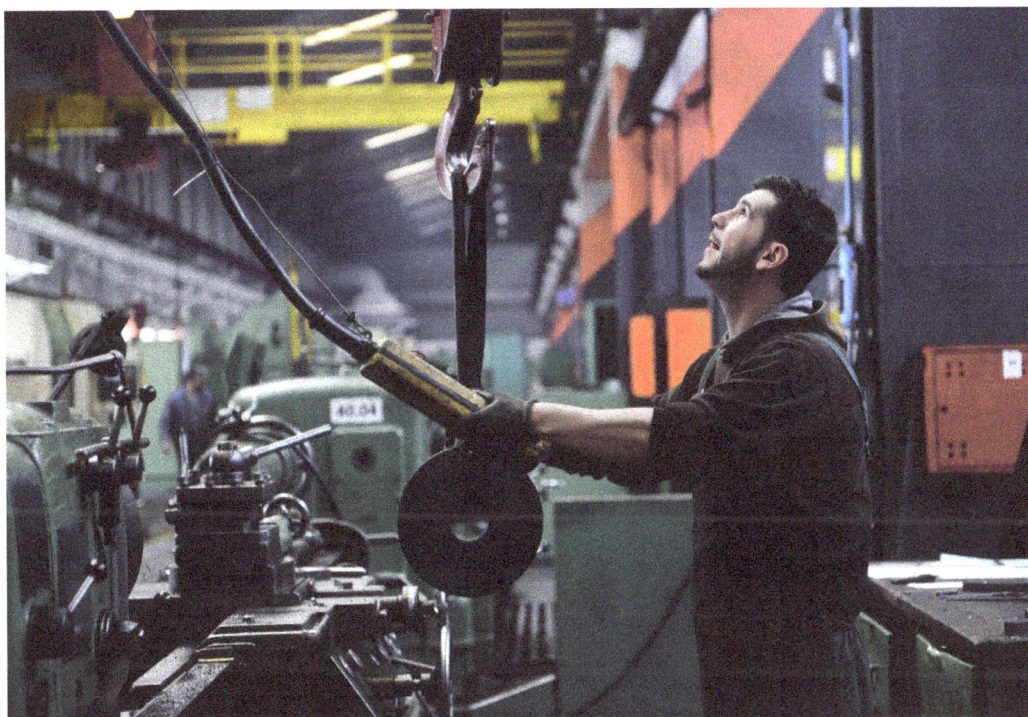

THIS CHAPTER PROVIDES:

- A description of the three components of loading.
- Descriptions of the body's response to loading.
- A brief overview of simple biomechanics.

How to cite this book chapter:
Berlin, C and Adams C 2017 *Production Ergonomics: Designing Work Systems to Support Optimal Human Performance.* Pp. 49–64. London: Ubiquity Press. DOI: https://doi.org/10.5334/bbe.c. License: CC-BY 4.0

WHY DO I NEED TO KNOW THIS AS AN ENGINEER?

This chapter will help you understand how much physical loading is acceptable in the workplaces you design. Based on the knowledge you have acquired in Chapter 2 (Basic Anatomy and Physiology), you now have an understanding of how the body's locomotive structures allow movement and the handling of loads. In this chapter, we turn that anatomical and physiological knowledge into mechanical principles of loading, allowing you to identify, analyse and evaluate the greatest risks for physical injury in the workplace.

One of the great strengths of engineering is the ability to make simplifications in order to calculate how much loading the body is under. If you have limit values available, biomechanical calculations can tell you whether a chosen task, in terms of posture, forces and time, will push the human beyond his or her limits. These simplifications are the basis for most ergonomics evaluation methods, which are explained in Chapter 8.

When you can identify unhealthy physical loading based on principles, you can reason your way into better decisions when choosing design solutions for the workplace.

WHICH ROLES BENEFIT FROM THIS KNOWLEDGE?

For the *system performance improver* and *work environment/safety specialist* striving to identify improvement potentials in a workplace, the previous chapter's anatomical and physiological knowledge may be overwhelming to keep in mind and difficult to separate into analytical components in order to look for risks in a structured way – therefore, this chapter provides an intermediate step along the way to the ergonomics evaluation methods by showing how the body's reactions to loading can be simplified into some main components that can be systematically observed and later targeted in improvements.

3.1. The components of physical loading

As you learned in Chapter 2, Basic Anatomy and Physiology, the body's tissues work together to withstand many different types of biomechanical loading. Exceeding the body's physical ability to handle these loads results in pain and physical injury, which can be either sudden or chronic. But if we regard the problem from an engineering perspective, we need concepts and methods to identify what exactly makes physical loading a risk.

To make this possible, we adopt the view that:

$$\textit{Physical Loading} = \textit{posture} \times \textit{forces} \times \textit{time}$$

Body posture demands that the body's muscles actively work to maintain a position, which is a form of *internal loading*. The posture aspect includes how internal forces are distributed across the different parts of the body (for example, lifting something off the ground with a straightened back engages mostly the leg muscles which are large and strong, while lifting the same object with a bent back loads the upper torso which has smaller, weaker muscles).

External loading occurs as a result of handling weights, e.g. by pushing, pulling, lifting, pressing or dragging something. Generally, when force is counted as a component of loading, we are mainly referring to external loading. In some biomechanical analyses, the weights of the human's own body parts are sometimes also considered a load, especially if gravity influences the chosen posture.

Finally, time factors describe how long, how often or how frequently the body's structures are loaded. Since you now know that the muscles and tissues can work for a limited time until they are fatigued and need to rest, the level of risk depends on whether the exposure is suitable for strength- or endurance-type body structures. The time component most frequently focuses on repetitiveness, which is considered a major health risk because the body's structures are not allowed enough recovery between loadings.

3.2. Posture

Posture denotes how the body is aligned and positioned, especially in states of activity. A posture can occur as a result of consciously choosing how to position the body, or less voluntarily as a result of adapting to available space, tool sizes, visual demands, pain, etc. Posture may be influenced by the contextual factors in Table 3.1.

Good and bad posture

There is a conception of "good" and "bad" posture, stemming from societal norms about keeping the body upright, symmetrical and well aligned. From a work design perspective, good posture is more than keeping your head upright and your back straight – it also includes strong hand postures, equal weight balance between the legs, and deliberately handling external loads close to the centre of the body. As a useful, operative definition for engineering work, we can define good and bad posture as follows:

> **Good posture** is a position where the functional structures of the body are in the best possible position to exert high force or high-precision movements, as required by the work task (Figure 3.1). Indications of good posture are balance, symmetrical distribution of forces on the body parts, and skeletal (rather than muscular) loading.
>
> **Bad posture** is a position where body is in a weak position to perform physically demanding work. Bad posture puts the body tissues under extra, unnecessary physical load that does not contribute to the task at hand. Indicators of bad posture include positions at the outer range or movement (hyperflexion or hyperextension), asymmetry, imbalance between the legs, slumping, and forced muscular loading rather than skeletal loading.

As stated in Chapter 2, different parts of the body are specialized for different types of movement and loading. For example, the back and legs are excellent at withstanding heavy loads for a long time

Table 3.1: Factors that may influence body posture.

SPACE	Humans are good at adapting body posture to existing preconditions in order to fulfil a task. This may often involve twisting or turning the body in order to reach, fit into an inconvenient space or avoid touching the surface of materials (example: so as not to scratch the paint job of a car). Therefore, it is necessary to determine how much working space around the task will be enough to avoid unnecessary loading, and whether to design for a minimum amount of space or with safety margins. A related aspect is to consider whether the available space will suit all body types and sizes[1].
VISION	An important prerequisite for performing a task is often being able to see what we are doing. If the line of vision is blocked or inconvenient, a human will often move the head, neck or upper torso to improve the line of sight, often bending or twisting. Therefore, visual demands can certainly influence posture. Also, insufficient lighting may have a similar effect even when the line of sight is acceptable, since it may still lead to bending closer to see controls, screen interfaces or instructions. It is a wise safeguard to have a well-lit working environment, particularly to ensure the ability to see[2] written information for workers of all ages.
STRESS	A high pace of work or high mental load (demanding tasks or working under pressure to perform) can contribute to feelings of stress. Heightened stress levels often increase muscular tension in the body, leading to a persistent internal loading situation that is static and can lead to fatigue. In some cases, tension from stress leads to cramping up and discomfort or pain. Stress can result from the psychosocial environment, demands of the job, the task speed or perceived mismatch between the task and the human's abilities.
PROTECTIVE CLOTHING	Many environments and tasks demand that the workforce should wear protective gear and clothing – sometimes to protect the human from extreme temperatures, glare, hazardous materials, wetness or dirt (e.g. gloves, glasses, jackets, helmets or visors), and sometimes to protect sensitive products or the environment from humans (e.g. hygiene masks and gloves). From a loading perspective, it is important to consider the additional postural load that these safety measures can bring about. For example, a helmet or visor may be heavy or warm, resulting in extra muscular effort and heat. Another example is that wearing gloves can often reduce surface friction and the sense of touch, leading to compensation with higher grip forces or clumsy use of hand tools.
	Finally, it is worthwhile to consider that protective clothing can impede both movement and vision.

if loaded in their axial direction, while the hands are highly flexible and responsive instruments of precision work rather than strength. Granted, with training some people are able to increase their force exertion in the hands or the precision of their back and leg movements, but it is generally reasonable to design tasks and workplaces so that they cater to what the body segments are naturally best at.

Causes and consequences of bad posture

Bad posture is often accompanied by initial warning signals in the form of tension, discomfort or pain. It often results from unawareness, ignoring signs of pain or discomfort, or underestimating the impact of low-level long-term loading. There is a conception that there are several ergonomics pitfalls

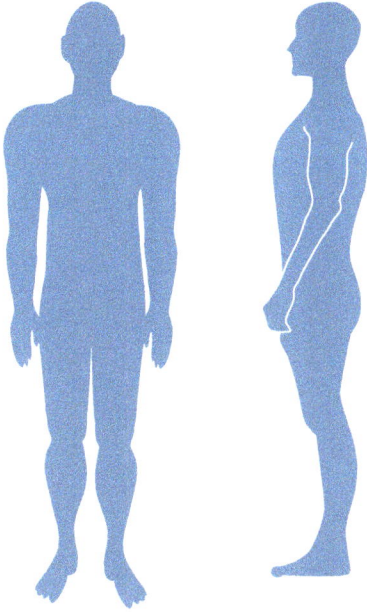

Good Body Posture

- Feet firmly planted on the ground
- Knees directly above the middle of the ankle joints
- Hips directly above the knees
- Shoulders squarely above the hips
- Head and neck held in a way that aligns the ear directly over the shoulders

Figure 3.1: Characteristics of good body posture (for the purpose of being ready for additional loading) Illustration by C. Berlin.

or typical scenarios that people often brush off as "not so bad" or just a minor inconvenience, but which may lead to risk for injury. These include:

- stretching to reach
- repeated heavy lifting
- lifting large, bulky, awkwardly shaped objects alone
- high pinch forces
- handling sharp, hot or cold objects
- working with hands above shoulders
- long periods of work holding the same body posture

As mentioned earlier, additional demands (such as seeing, avoiding touching surfaces, psychosocial issues, or compensating for protective gear with posture or force) may be part of these ergonomics pitfalls. Observable work behaviours include bending, pushing, pulling, lifting, hand twisting, unbalanced standing or sitting, and repetitive actions.

Some postures themselves can cause static loading on the body, meaning that forces or torques are applied for so long on the engaged body parts that they are not given sufficient rest. This can lead to fatigue, decreased force/precision performance, and compensation recruitment of extra muscle fibres. In many cases, static loading leads to constant tension in the muscles which can lead to tiredness, discomfort and cramping or even headaches. Such static postures and loading situations include:

- bending the back forwards or sideways
- holding loads in the hands

- stretching the arms out to the sides or raising them above the shoulders
- putting weight on one leg, while the other works (e.g. a pedal)
- standing in one place for long periods
- sitting in one place for long periods (e.g. computer work or driving a car)
- pushing and pulling very heavy objects
- tilting the head forwards or backwards at the extreme end of motion to see
- raising the shoulders

Measuring posture

How, then, can we determine if a posture in itself is harmful? A good rule of thumb is that if a posture is held near the outer range of motion, it is probably not a good position for taking on external forces. For many ergonomics evaluation methods, posture is defined in terms of joint angles between body segments. A "neutral" posture is considered the least amount of loading, and resembles a relaxed, standing, symmetrical body position with the arms hanging along the sides of the body (Figure 3.2). Deviating from this relaxed, standing posture is considered an increase in risk for harmful loading.

For situations where work postures are being observed or assessed manually, rough estimates (based on the expertise of a trained eye) of the joint angles are often sufficient, but for analyses that require more precise values for joint angles, the following measurement methods exist:

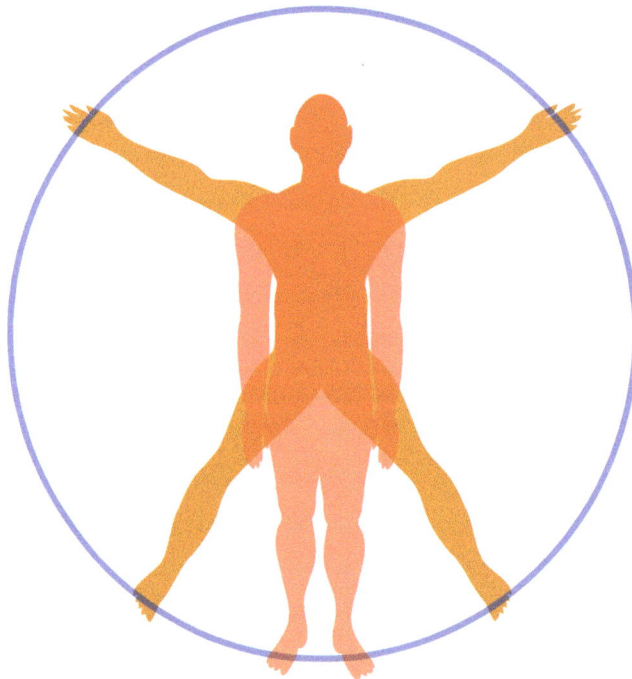

Figure 3.2: The basic "neutral" posture (red), typically considered as the lowest risk for harmful loading, and a near-maximal deviation of limbs (orange) from that neutral position, generating biomechanical torque on most joints. Bending and twisting also lead to deviation from the "ideal" starting posture.
Illustration by C. Berlin.

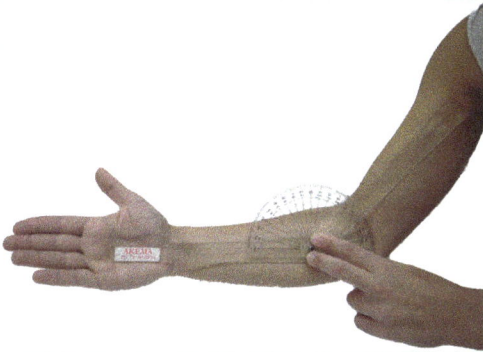

Goniometers[a] are graded tools used to measure angles between body segments.

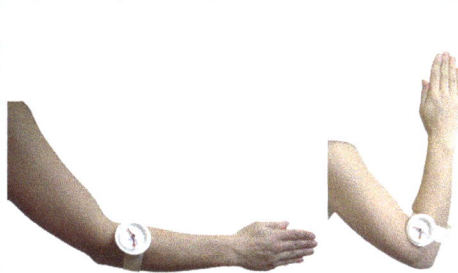

Inclinometers are electronic devices that can be mounted to body segments to continually log and measure the "leaning" or inline of body parts in different positions.

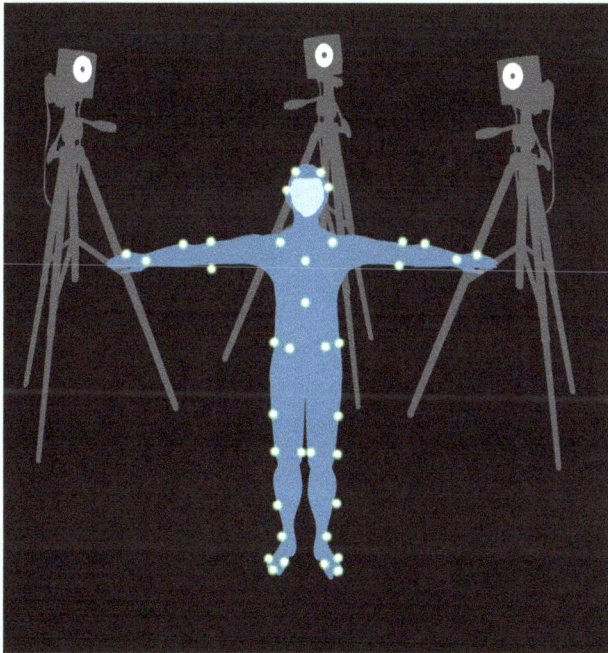

Motion capture has been used primarily by the film and games industries to record the motions of a real human being performing movements. Early motion-capture technologies often involved attaching electrodes with wires to different body segments. Today, this is usually done by one of two ways: 1) *visual motion capture*, where the person is strapped with reflective visual markers, and recording of the person's movements is from all angles, using several different cameras simultaneously, and 2) *wireless motion capture*, where digitally connected sensors with inclinometers are oriented to a 3D coordinate system in a computer and wirelessly transmit human movement as the recording progresses. The result in both cases is a file that registers how the markers move relative to each other in a 3D space. This recording can be imported into ergonomics evaluation software, and joint angles and various risk levels can be easily deduced from…

...**Manikins**[b], which are 3D representations of humans in a 3D CAD environment that can be posed and made to move. Since manikins are frequently used to evaluate posture and ergonomics, exact joint angles are generally easy to obtain from the software that the manikin is used in.

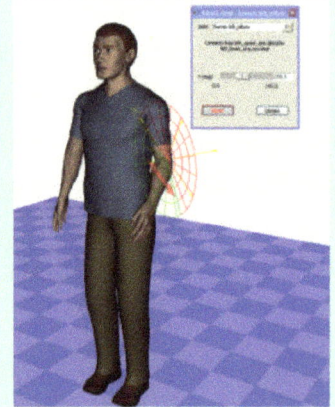

[a] Images by C. Adams (Goniometers and Inclinometers) and C. Berlin (Motion Capture).
[b] © 2016 Siemens Product Lifecycle Management Software Inc. Reprinted with permission.

3.3. Force

Force in itself is only a risk if it exceeds the limit loading values of the body's structures. Some of these limits are determined by materials science values for body tissues, but a certain degree of ability to handle large forces actively can be influenced by training, health status, nutrition levels and genetic preconditions.

In static mechanics, a force is traditionally thought of as a vector arrow with a certain magnitude and direction, acting on a point. But to study the impact of real-life loading forces, we need a more nuanced vocabulary to do justice to forces. Table 3.2 shows some different terms by which we can characterize force.

Table 3.2: Terminology concerning forces.

MASS	The inert weight of objects that are not in motion, expressed in kg or lb.
DYNAMIC FORCES	Forces that have variation in magnitude and direction, engaging different muscle groups and leading to aerobic (oxygen-based) processes in the muscles.
STATIC FORCES	Forces that affect a limited muscle group for a sustained period of time, allowing little or no rest and recovery. This leads to discomfort, fatigue and anaerobic processes (production of painful lactic acid) in the muscles.
REPETITIVE FORCES	A special case of static loading, these are forces that are short in duration, but occur so frequently that the muscles are not able to relax in between loadings, meaning that their overall load is equivalent to a static force.
EXTERNAL FORCES	External forces often occur as a result of handling objects by pushing, pulling, lifting, lowering and carrying.
INTERNAL FORCES	Internal forces arise when the body's muscles strive to maintain a posture, either as a reaction to external loads or because of higher internal pressure at the extreme ends of our range of motion.

Figure 3.3: Force gauge for measuring push and pull force. The readout is often given in *N*.

Measuring force

As with posture, rough estimates can go a long way, but it is often necessary to get a value on the force being applied to judge its risk impact. A very rough yet effective method is to use weigh scales (such as bathroom scales or luggage scales with a hook, Figure 3.3) to measure push and pull forces expressed in kilograms or pounds. This can then be roughly approximated into force expressed in Newtons by multiplying the gravitational factor 9,.82 m/s^2. For more exact force measurements, force gauges for measuring pushing or pulling motions can be used (see Figure 3.3).

3.4. Time

Time factors can significantly influence the occurrence of work-related MSDs and make a seemingly small and harmless load into a risk for long-term injury due to wearing out the body. Primarily, it is important that loading from tasks must be suitable for the body tissues that are engaged and that they are allowed sufficient rest and recovery between exposures. Exposure can be defined as the time duration that the body's structures are actively engaged in order to perform a task, usually in order to sustain a force or torque.

Table 3.3: Terminology of time exposures.

REPETITIVENESS	Repetitiveness, also known as "monotonous work", is thought of as the potentially most harmful time exposure factor. Generally, the magnitude of force is not the problem with repeated loading; the lack of recovery is. Since repeated motions affecting the same muscle groups lead to little or no time for rest, this type of exposure is considered equivalent to static loading.
	Definitions in scientific literature vary regarding limit values for repetitiveness, but many definitions count the number of "same" actions that occur every 30 seconds (Zandin, 2001).
	Repetitiveness can be either measured as the speed at which the operator carries out the tasks, or it can be measured in terms of the number of movements or posture changes per shift.
FREQUENCY	Frequency designates the number of occurrences per time unit that a muscularly similar action occurs. Repetitiveness is often expressed in terms of frequency.
CYCLE TIME	The inverse of frequency is the cycle time, i.e. time duration per completed motion or task.
ENDURANCE TIME	The period of time before fatigue sets in; until that time, the body tissues can tolerate constant or repeated loading and still function to a satisfactory level of speed, precision and/or strength.
FATIGUE	The state where musculo-skeletal structures are loaded to the extent that they can no longer exert sufficient force, speed, precision or motion range anymore. At this stage, rest must begin to achieve recovery and rebuild safety margins against physical injury.
RECOVERY	The state where musculo-skeletal structures are free of discomfort, tiredness and pain related to exposure, and are once again ready to take on loading.
RESUMPTION TIME	The time it takes between reaching the stage of fatigue and when the worker feels ready to resume the activity or task.
CUMULATIVE LOADING	Cumulative loading is the notion that load exposures add up over time, and that some injury risks are difficult to identify unless the loading is considered over different time perspectives. This is especially true if there is routinely insufficient rest or recovery. For example, load risks may not be evident when studying cycle times of ~30 seconds, but may emerge if the loading is considered over an hour, over a shift, over a day, a month... all the way up to an entire working life. For some manual labour professions, certain types of loading may have a significant physical impact over the course of a working life.
VARIATION	The main remedy against harmful time exposures is to introduce variation – this means doing a variety of different tasks after each other to avoid repetitive motions. It is believed that even if muscular activity remains high, spreading out the loading on different body structures gives the different muscle groups a chance for relative rest and recovery.

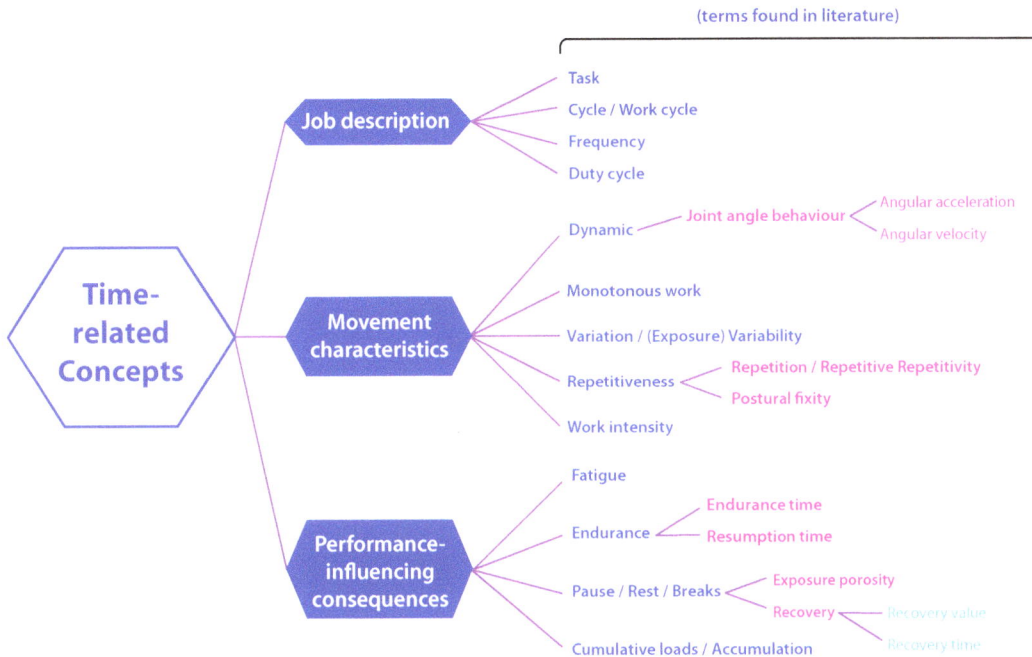

Figure 3.4: A hierarchy of time-related factors that can be used to describe production assembly work. (Adapted from Berlin & Kajaks, 2010).

llustration by C. Berlin, based on Berlin & Kajaks (2010).

3.5. Interaction of posture, forces and time

It is important to remember that the interaction between posture, force and time may sometimes increase or decrease the total risk (increased probability and severity of injuries) considerably. It is for instance not necessarily true that lifting heavy weights is always a risk; this is acceptable as long as it is done infrequently (to ensure recovery) and with good posture. In contrast, small, persistent loadings over a long time period can be much more harmful than they seem, because weak structures that are constantly nearing fatigue can "drag along" neighbouring body structures into compensating with muscular tension.

Sometimes, the nature of the task can also influence whether loading is harmful or not. Often it is a question of whether the three components are of a suitable magnitude. You learned earlier that high-precision work with the hands is not good to combine with maximum force. It then follows that different hand postures or grips are ideal for high-precision or high-power work respectively. However, some postures alone will raise the risk greatly – working with highly flexed or extended wrists is both harmful and ineffective, since most extreme-range postures lead to nerve and tendon entrapment and provide a weak position for transferring force.

To describe the risk levels of these factors combined, the *cube model*[3] (Sperling et al., 1993) gives each of the loading components three criteria levels of severity (where 1 = low risk and 3 = high risk) showing which combinations may result in harmful loading or injuries. Figure 3.5 shows that a high level on just one out of three components may be acceptable as long as the value is lower than 6,

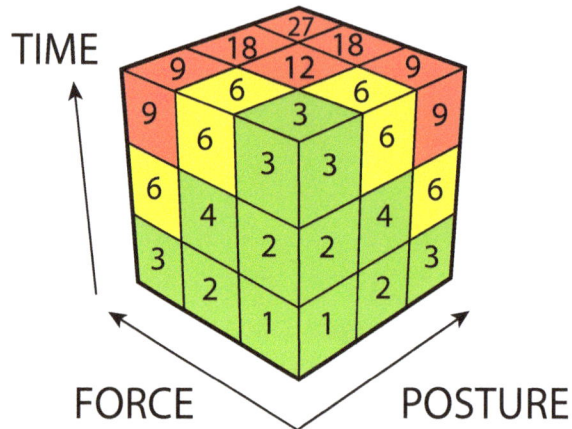

Figure 3.5: The cube model, showing how different combinations of posture, forces and time result in different risk levels. (Adapted from Sperling et al., 1993).

Illustration by C. Berlin, based on Sperling et al. (1993).

(green zone), while a combined level that is higher lands in the yellow (6 and above, but under 9) or red zones (9 and above), indicating that the load must be reduced (red) or at least investigated (yellow).

3.6. Other factors influencing physical loading

Some additional non-anatomical factors in a work environment may affect physical loading in a way that engages all three loading components or combinations of two of them.

Table 3.4: Factors that may have influence on body posture.

VIBRATIONS	Vibrations are a special form of loading; different body tissues have different resonance frequencies and therefore have different sensitivity levels to vibration forces. The body spends the entire times it is exposed to vibration compensating muscularly for small external forces that act in opposite directions on the body. This compensation tension can lead to sustained strain, over time resulting in cumulative trauma disorders.
ENVIRONMENTAL FACTORS	Cold, heat and humidity can affect contact comfort, grip friction, forces required, avoidance of burning or chill, etc.
NON-RIGID MATERIALS HANDLING	Rubber, fabric, cables, etc. are often large, floppy, sometimes elastic and difficult to manage in a consistent standard way. Extra force exertion may be necessary due to friction, dragging material on the floor, elastic behaviours and entanglement. It is also difficult to measure the forces required, both during carrying and during assembly.
HIGH-PRECISION WORK	High-precision work increases demands on performance and requires extra suitable working conditions and postures in order to be executed efficiently and with high quality.

THE "CINDERELLA HYPOTHESIS"

Hägg (1991) proposed a theory of cumulative loading known as the "Cinderella Hypothesis", named after the fairy tale character that was always "first to rise and last to go to bed". This theory aims to explain why there is a risk for injury even when humans perform tasks with low forces over prolonged time duration.

The basic idea is that when a motion occurs and the muscle contracts, certain low-threshold (weak) motor units are recruited first (with other stronger ones successively joining in as the motion continues) and are deactivated last. This means that when motions are repetitive, some muscle fibres run a greater risk of injury, because even though the motion stops briefly, the first recruited fibres remain constantly activated at low loads, meaning that there is no recovery. This leads to fatigue, pain and possibly cumulative strain injuries. This theory helps to explain low-load musculo-skeletal problems in the neck, shoulders and wrists, such as mouse arm and writing cramps.

3.7. Biomechanics

This book only gives a very brief overview of biomechanics, for the purpose of introducing you to the basic assumptions and simplifications behind many ergonomics evaluation methods. It is not intended to be extensive, so it will only bring up some very simple examples. To read more extensively on the subject, please consult a dedicated textbook of biomechanics, such as Knudson (2007).

3.8. Applying mechanics to the human body

The human body is made up of many different tissues (bone, muscles, nerves, ligaments, etc.) that all have different mechanical properties, for example limit values for loading and strain. Furthermore, they can move in a three-dimensional range of motion during loading. However, it is possible to simplify calculations of how much the body can be loaded using simple laws of mechanics, and by considering motions in a simplified way: by studying the forces and torques acting on the body at different "before" and "after" positions in a two-dimensional plane. The biomechanical calculation is often made with loading on a specific body part as reference, and generally, there is a limit value for how much force or torque that body part can safely withstand. However, since exact biomechanical calculations are extremely complicated, a simplified equation must be built on many assumptions. To be able to trust the simplification of the physics acting on the body, the limit value for how much a body part can be loaded should be calculated conservatively – in other words, the safety of the body is ensured by considering its weakest link.

Some basic assumptions (or simplifications) made in biomechanical calculations are that:

• Skeletal bones are considered as rigid bodies (no plasticity).
• Joint motions are considered in one direction.

- There is no friction in the joints.
- Torque is considered to affect only one muscle or muscle group in one direction.
- There are no antagonistic muscle forces.
- The mass of the body segment is calculated as a percentage of total body weight.
- Body weight and measures for centre of gravity are taken from anthropometric literature data.

Stress, strain and trauma

In a biomechanical sense, stress is defined as potentially harmful loading. Stress is usually the result of forces or torques acting on the body structures, up to the point of strain, meaning that the structures experience deformation as a result of the loading. This in turn goes to the point of trauma, which means that the structures fail or break. Every tissue in the human body has its limit value of stress that it can withstand before failure. As long as loading is beneath that value, the structure is safe, but above it, risk for injury is present.

Study questions

Warm-up:

Q3.1)   What is the difference between internal and external loading?

Q3.2)   Name some causes of bad posture that may arise from the work task and work environment.

Q3.3)   What is the difference between dynamic and static loading?

Look around you:

Q3.4)   Find some videos online (for example on YouTube.com) showing physical assembly work tasks; can you use the posture/forces/time triad to identify risks for unhealthy physical loading?

Q3.5)   Reflect on your own working life as a student, engineer or the like. What are the typical postures, forces and time frequencies of exposure that occur in your daily life? Are you at risk for unhealthy loading?

Connect this knowledge to an improvement project

- When observing physical work, look for recurring posture-, force- and time-related risk occurrences.
- Try to identify the root cause – in the task or environment – that may cause or contribute to the previously mentioned risk exposures.

• Ask operators why certain behaviours are adopted. If there is a known reason, this will perpetuate the risky behaviour and should be addressed. What function does the answer to that "why" fill?

Connection to other topics in this book:

• Some ergonomics evaluation methods (Chapter 8) specifically target one or more of the risk factors of posture, forces and/or time. When choosing a method, consider that:
  • Many methods are purely posture-based (Chapter 8), meaning that they may exaggerate the severity of the posture if it is not frequently occurring.
  • Time-related evaluation methods are not commonly covered; at least not with observation-based methods that you can perform on-site. Usually some assumptions are needed.
  • Force-related evaluation exists and is well backed up scientifically, but many of these guideline rules are limited to a specific population (for example, by being valid mostly for men), so it is worthwhile to be aware of how anthropometry (Chapter 4) dictates how relevant these methods are.

Summary

• Physical loading is a combination of posture, force and time.
• Posture dictates how the body is aligned and positioned and is influenced by space, vision, stress and protective clothing.
• To maintain a certain posture the muscles must actively work; this is a form of internal loading.
• A good functional working posture is one in which the body is balanced, forces are symmetrically distributed over the body and external loads are held close to the body while both feet are firmly planted on the floor.
• Bad demanding postures, where there is an imbalance between the legs, extensive muscular loading and movements at the outer range should be avoided where possible.
• A neutral posture where the body is relaxed and symmetrical with the arms close to the body is considered to involve the least amount of loading.
• Static loading when forces and torques are applied for prolonged periods of time without sufficient rest should be minimized.
• Excessive static loading can lead to fatigue, decreased performance levels, constant tension in the muscles and discomfort.
• Forces are only an injury risk when they exceed the loading value of the body's structures.
• Static forces affect a limited muscle group for a sustained period of time with little or no rest and recovery.
• Dynamic forces have variation in magnitude and direction, engaging different muscle groups.

- Forces can be both internal and external.
- Time factors describe how long, how often or how frequently the body's structures are loaded.
- Repetitiveness is one of the most harmful time factors, when repeated motions affect the same muscle groups with no time for rest.
- Fatigue is the state at which musculo-skeletal structures are loaded so much that they can no longer exert sufficient force, speed or precision.
- Rest is key to enable the body to recover from fatigue so it can function normally again.
- Variation of body postures and applied loads coupled with sufficient recovery time is very important during work.
- Applying principles of mechanics to the human body is known as biomechanics and can be used to calculate what loads the body can withstand.

Notes

1. We explain how to consider different body sizes in Chapter 4: Anthropometry.
2. We explain how to consider vision and lighting in Chapter 12: Environmental Factors.
3. Although the Cube Model was originally developed to evaluate hand/wrist loading when using hand-held tools, the logic of interaction between these loading risk components is applicable for the entire body.

3.9. References

Berlin, C. & Kajaks, T. (2010). Time-related ergonomics evaluation for DHMs: a literature review. *International Journal of Human Factors Modelling and Simulation,* 1(4); 356–379.

Hägg, G. (1991). Static workloads and occupational myalgia — a new explanation model. In Anderson, D.J. Hobart, and J.V. Danhoff (Eds.), *Electromyographical Kinesiology,* 141–144.

Knudson, D. (2007). *Fundamentals of Biomechanics.* New York, NY: Springer.

Sperling, L., Dahlman, S., Wikström, L., Kilbom, Å. & Kadefors, R. (1993). A cube model for the classification of work with hand tools and the formulation of functional requirements. *Applied Ergonomics,* 24 (3); 212–220.

Zandin, K. B. (2001). *Maynard's Industrial Engineering Handbook,* 5th edition. [Online] New York: McGraw Hill. Available from: http://accessengineeringlibrary.com/browse/maynards-industrial-engineering-handbook-fifth-edition/p2000a1fc99706.9001 [Accessed 16 January 2014].

CHAPTER 4

Anthropometry

THIS CHAPTER PROVIDES:

- Theory on statistical variation in human populations.
- Design procedures for selecting critical users to design for.

How to cite this book chapter:
Berlin, C and Adams C 2017 *Production Ergonomics: Designing Work Systems to Support Optimal Human Performance.* Pp. 65–82. London: Ubiquity Press. DOI: https://doi.org/10.5334/bbe.d. License: CC-BY 4.0

WHY DO I NEED TO KNOW THIS AS AN ENGINEER?

Perhaps you have heard of something called the "average person". If you take anything with you from this book, let it be the knowledge that the average person *does not exist*. At least, the average person is not somebody you can or should design workplaces or equipment for. While it is possible to have average height, average grip strength or average weight, there are too many possible individual combinations of biological variations to design for any "standard" person, and sometimes the "middle" or mean is not where the statistical majority of people are found. Instead, most workplaces need to be designed for a range of needs from a population of people, ranging from small to large sizes in a number of different ways. This is the best way to accommodate as many system users as possible.

In other words, this chapter transfers the focus from the needs and capabilities of the *individual*, to the needs of the collective – so that our engineering solutions become useful for a *population*. Anthropometry is the study of statistical variation of human body dimensions and its implications on design. This concerns everything from workplaces, tools, vehicles and medical packaging to clothing. For an engineer, a helpful design input is to know the measurements of "critical users" whose specific needs must be met by the workplace design dimensions in order for them to be able to work in the most effective, productive and risk-free way.

WHICH ROLES BENEFIT FROM THIS KNOWLEDGE?

The *system performance improver* gains an understanding for the range of worker body sizes, strengths, etc. that the workplace and its equipment needs to be dimensioned for, especially for future recruits or an aging population. The *work environment/safety specialist* will be able to identify workplace risks and improvement potentials that are caused by a mismatch between worker size, strength, etc. and available equipment. The *purchaser* will be able to better understand the business sense in investing money in adjustable solutions that fit more workers (in spite of the perceived higher cost at the purchase stage), but may require a business case example and consideration of benefits for the whole workforce to be convinced. The *sustainability agent* will be able to connect ergonomics very clearly to demographic developments and align the design of the workplace to social sustainability concerns, such as readying the workplace for future workers.

4.1. Designing for the human

It is important for work environments to be designed according to the characteristics of the human body. Anthropometry is the branch of science that deals with human body measurements; its name comes from Greek, where *Antropos* means human and *Metrikos* means measurement. As a discipline, anthropometry dates back for centuries with many people taking an interest in the proportions of the human body.

We have previously discussed how the human body reacts to loading, and in Chapter 8 we introduce a number of tools and methods to evaluate workplace situations and identify areas for improvement, so now theoretical knowledge based on the physical characteristics of the human body will be discussed to aid in making improvements and redesigns.

There is a large variation in body size from one person to the next, with people having unique proportions across each body segment. There is significant variation in body size between different populations, genders and nationalities, which makes the design of equipment and workstations challenging, but this must be taken into consideration, especially when designing for an international environment. For example, a piece of equipment designed to fit 90% of Americans may suit 90% of Germans, but only 65% of Italians and 45% of Japanese, if we look at the size ranges in those local populations. However, populations also change over time, reflecting the effects of migration and genetic developments, so the best bet is to design your work equipment or environment for a range of populations and to use as recent databases as possible.

When it comes to designing for the human, the "one size fits all" approach rarely provides satisfaction for all involved. Just as the clothing industry takes variation into consideration by providing a range of sizes to meet everyone's diverse needs, a number of considerations must be made to enable a diverse range of people to all use one workstation setup. In reality, there are very few work environments that are custom designed and tailored to one specific individual (Formula 1 cars are one of the rare exceptions). While individually designed workstations would probably promote healthier working practices, they would be extremely expensive and impractical. Instead, it is necessary to

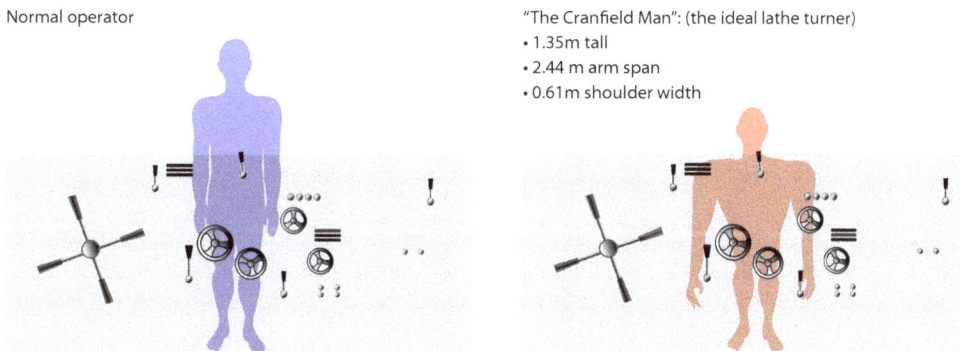

Figure 4.1: The "Cranfield man" on the right illustrates the mismatch between real operator measurements and a machine's controls (Eastman Kodak Company, 1983).

Image by C. Berlin, inspired by Eastman Kodak Company (1983) and Kroemer (2010).

select appropriate sizes for different aspects of the design, taking into consideration the variations in body measurements across populations, so a solution for the majority of the population is achieved. Anthropometric data plays a key part in this process of optimizing a design to maximise its use and value for the greatest number of users.

A study at the Cranfield Technology Institute highlighted the issue of humans not being considered during the design of a lathe, when they calculated that the ideal operator for one such machine would be 1.35 m tall, with an arm span of 2.44 m and a shoulder width of 0.61 m in order to operate the machine and turn the handles (Singleton, 1964).

4.2. Terminology

Glossary of statistical terms

NORMAL DISTRIBUTION	Also known as Gaussian distribution – when a set of data measurements follows a bell curve with a high frequency of occurrences around the mean and few values at the extremes.
PERCENTILE	Percentage point on the measurement distribution; the cutoff point in a population at which that percentage has a certain characteristic limit measurement, and the rest do not.
CORRELATION	When a strong relationship exists between different body measurements; i.e., if one measurement moves toward an extreme, then so does another. This relationship can be determined using statistics. The value r is used to indicate if the correlation between measures is positive (both measures move in the same direction) or negative (when one increases, the other decreases). The value r = 1 indicates a maximally positive correlation, r = 0 is no correlation, and r = –1 indicates a negative correlation.
POPULATION	Term to describe a particular group of people of interest who have been selected due to a certain characteristic, e.g. age, nationality or gender.
VARIATION	Difference within a particular body measurement across populations.
BIVARIATE	Concerning the design of solutions where two measurement variables are taken into account simultaneously.
MULTIVARIATE	Concerning the design of solutions where several different measurement variables are taken into consideration simultaneously.

4.3. Static (structural) measurements

There are two different types of anthropometric measurements: static and dynamic.

Static measurements describe dimensions and distances that are taken while people are in a defined, unmoving position. Measurement points known as *landmarks* (see Figure 4.2) are positioned over the human body and measurements are taken in a straight line from one landmark to another.

Static measurements are very specifically defined and include stature, eye height, sitting height, buttock-to-knee length, etc., as shown in Figure 4.3.

While these dimensions are relatively easy to obtain, they have limited value when designing workplaces since the body rarely adopts such predetermined specific positions during real work.

Abdominal point, anterior: The most protruding point of the relaxed abdomen on a sitting participant.

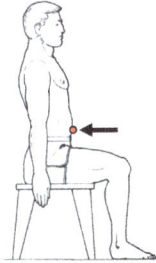

Acromion, right and left: The point of intersection of the lateral border of the acromial process and a line running down the middle of the shoulder from the neck to the tip of the shoulder.

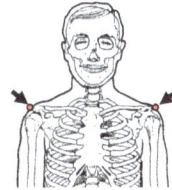

Figure 4.2: Two examples of definitions of landmarks for static measurements (Slightly modified figure from Gordon et al. 2014 p. 20).

Image permissions for Figures 4.2 and 4.3 granted by U.S. Army Natick Research, Development, & Engineering Center. The images have been slightly modified from the originals (in 4.2 surrounding table lines are removed, and in 4.3 the figure labels are moved to the side).

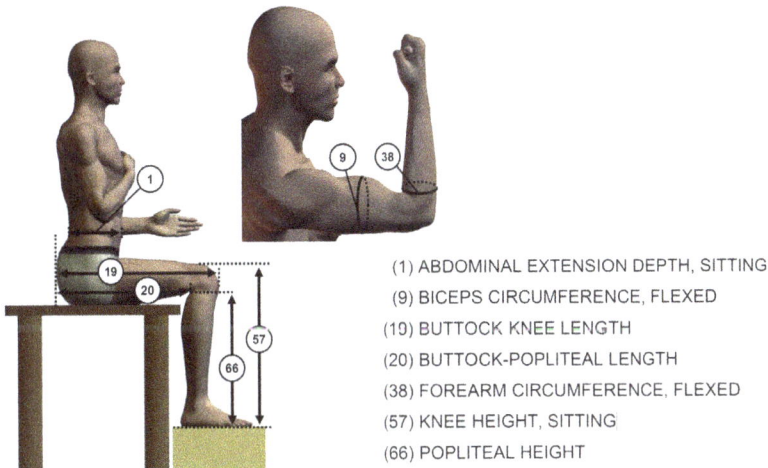

(1) ABDOMINAL EXTENSION DEPTH, SITTING
(9) BICEPS CIRCUMFERENCE, FLEXED
(19) BUTTOCK KNEE LENGTH
(20) BUTTOCK-POPLITEAL LENGTH
(38) FOREARM CIRCUMFERENCE, FLEXED
(57) KNEE HEIGHT, SITTING
(66) POPLITEAL HEIGHT

Figure 4.3: Static measurements (Slightly modified figure from Gordon et al. 2014 p. 402).

4.4. Dynamic (functional) measurements

Functional measurements concern dynamic positions, providing information about the necessary space required to carry out certain movements. While these measurements are more relevant to the design of workspaces, the data available in databases is usually very specific to particular work scenarios, so care should be taken when basing designs on such measurements. Examples of dynamic measures include ranges of reach (see Figure 4.3), clearance (how much space a person or body part takes up in relation to an object's boundaries, e.g. when passing through a doorway), strength measurement, etc. Obtaining and measuring functional measurements is more difficult than static measurements,

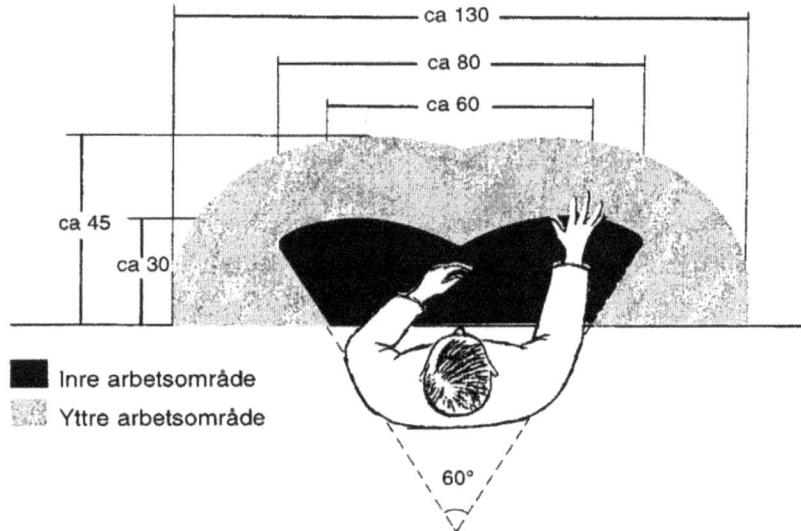

Figure 4.4: Dynamic measurements: reach distance design guidelines (Swedish Work Environment Authority, 1998).

since many measures involve multiple body actions and movements in concert. For example, dynamic reach may include bending towards an object as well as extending the arm. This makes the act of standardizing the measurement quite complex, and thus databases of such measures are not easily verified.

4.5. Normal distribution and percentiles

Figure 4.5 shows a typical distribution of anthropometric data for stature (height). The distribution follows a bell curve, which in statistical terms is known as a "normal" or Gaussian distribution. In such a distribution, the mean, median and mode values are the same.

This curve is almost symmetrical about the highest point, which is the mean (average) height and the most probable height to occur (given it has the highest frequency), so 50% of the population in question are shorter than the mean and the other 50% are taller. In contrast to the high frequency of people close to the mean height, there are few very tall or very short people, as can be seen from the two tails of the curve. To better understand what percentage of the population have a certain stature, the x-axis can be split into sections, where each section or division is known as a percentile. Percentiles can be calculated if both the mean and standard deviation of a group of measurements is known. If someone has 5th percentile stature, it means they are taller than 5% of the population, while someone with 95th percentile stature would be taller than 95% of the population (with only 5% of the population being taller). The concept of percentiles can be applied to any measurement of the human body, including non-visible measures such as hand strength. While a lot of anthropometric measurements can be approximated using a normal distribution curve, this is not the case for weight, depth, width and strength measurements. A person's percentile measurements are rarely consistent

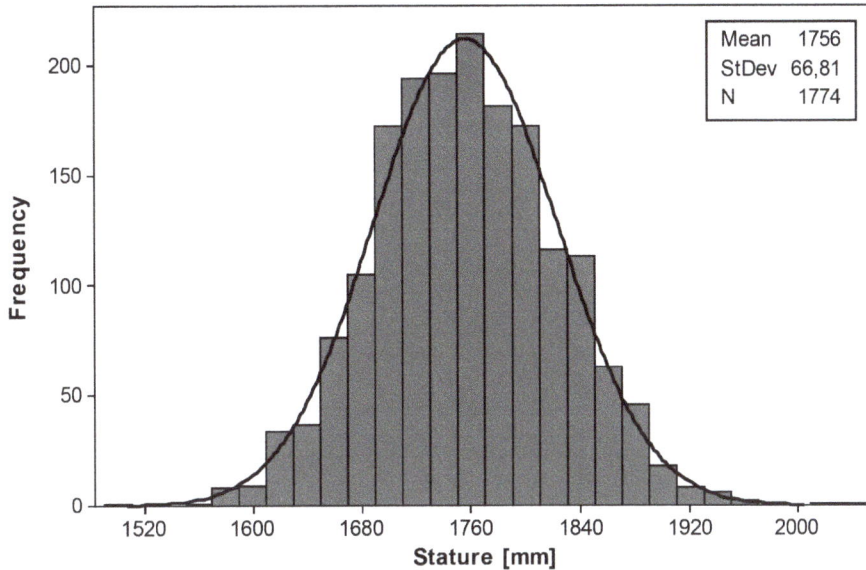

Figure 4.5: Normal distribution plot of stature (in Brolin, 2012; based on data from Gordon et al., 1989). Image reproduced with permission from E. Brolin. All rights reserved.

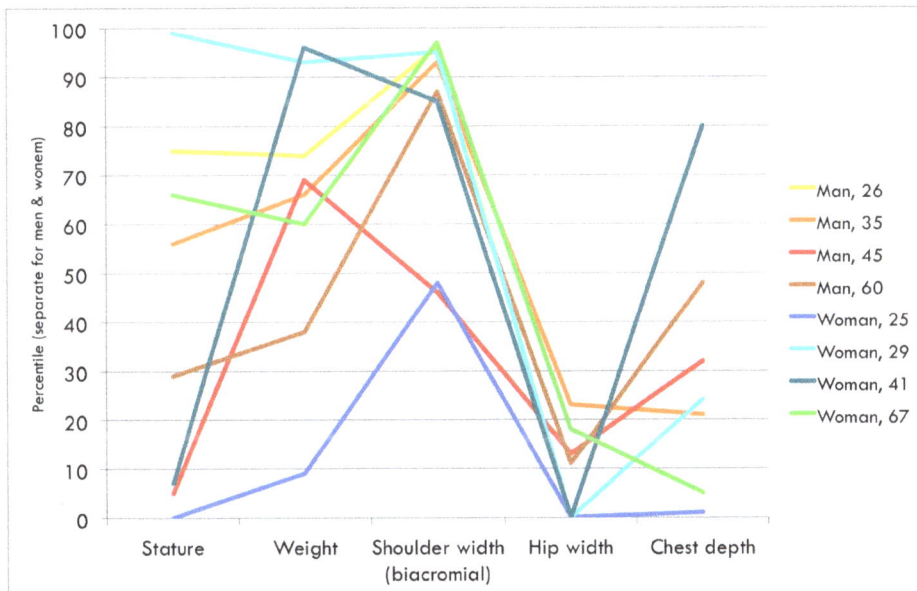

Figure 4.6: Variation in individuals' measurements – note that each individual person's combination of measurements fall into widely different percentile ranges!

Image by C. Berlin, based on data from Heinz et al. (2003) and Hanson et al. (2009).

across their entire body; while they may be 80th percentile in stature, they would be unlikely to be 80th percentile in all other measurements. As can be seen from the sample set of measurements in Figure 4.6, people with constant percentile values for a number of dimensions are rare, and therefore it is not meaningful to assume that a consistently "average" person exists or could be representative for the needs of a population.

Using the concept of percentiles, it is possible for designers to decide from the outset exactly which portion of the population they want their solution to be suitable for. In ergonomics and the design of workplaces, the extreme measurements are considered the most interesting, as these are the boundary attributes that could cause a design to be unsuitable and "not fit" the intended workforce. Generally when selecting which data to base designs on, one should ask the question:

"Who will be excluded from this solution if I select these measurements and what are the implications?"

However, care should also be taken, as designing for the extremes can mean that the solution is sub-optimal for the majority who aren't considered extreme.

4.6. Correlations

Some body measurements are closely related; for instance, eye height is, logically, closely connected to stature. However, this is not the case with all measurements; for example, head circumference shows no such relationship with stature. Statistically speaking it's possible to determine how strong a relationship exists between different sets of data using the Pearson correlation coefficient (r). This measure provides information about the level of dependency between two variables, giving a value between -1 and 1, where 1 is a perfect positive correlation, 0 is no correlation and -1 is a negative correlation (e.g. as one variable increases the other variable will decrease at the same rate. For example: the more time you spend at work, the less time you spend at home).

Generally where anthropometry is concerned, measurements need to demonstrate an *r* value of at least 0.7 for them to be considered correlated (Figure 4.7 and Figure 4.8).

It is valuable to understand how measurements across populations and body segments differ and how anthropometric data sets have come to be. Relationships between some measurements of the US air force showed that correlation exists between: stature and overhead reach, stature and wrist height, stature and sitting height, and stature and span. So while stature is the easiest measurement to obtain, it is not sufficient in many cases to use it as a predictor for other measurements. Care should be taken to generalise this information as it comes from a very specific group. Given that there is not a direct relationship between all measurements, it is not possible to add percentile values together.

4.7. Multivariate design

Typically it is not sufficient to only take one body measurement into account when designing workstations; rather, a number of different measurements are considered. This is known as multivariate design, and when the design of a solution only takes into account two measurements it is known as bivariate design.

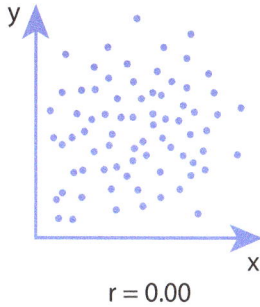

Figure 4.7: Uncorrelated measurements.   **Figure 4.8:** Correlated measurements.
Images by C. Berlin.

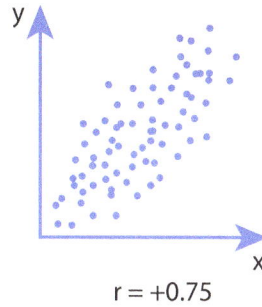

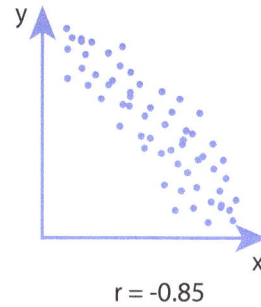

In cases where more than one measurement is used by the designer and a 5th percentile to 95th percentile approach is adopted, the reality is that the dataset actually excludes more than 10% of the population, as can be seen in Figure 4.8.

The design is in fact only suitable for those who fall within the squared area that only contains 82% of the population, thus excluding 18%. By adding a third measurement, a multivariate case is introduced and the percentage of the population accommodated by the design will be even less. This can be plotted on a 3D graph.

4.8. Variation

While almost every human body has the same "biomechanical layout", there is significant variation in body sizes and proportions between individuals. The main reasons for variation between anthropometric data are due to:

- data management
- intra-individual variations
- gender
- nationality
- age

Data management

The first reason for variability between measurements has nothing to do with physical variation between groups, but is actually due to poor data management. By not adopting standardized methods and utilizing illogical procedures while taking, analyzing and organizing measurements, errors can easily occur. This results in unusual measurements that don't accurately represent reality. Should you encounter any measurements which are significantly different from any other published data set then extra care should be taken when designing workplaces based on such numbers.

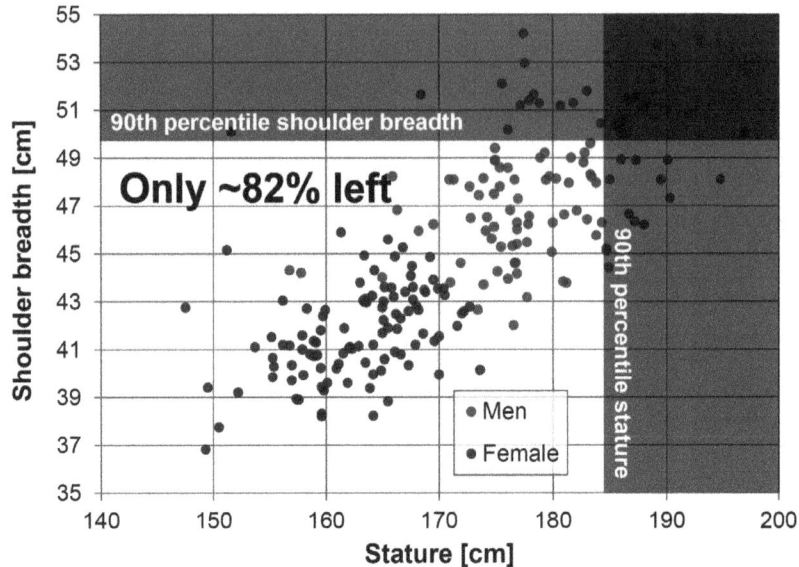

Figure 4.9: Bivariate frequency distribution of statue and weight – note how the 90th percentile principle of exclusion in two uncorrelated measurements ends up excluding about 18% of the population
(Brolin, 2013).

Intra-individual variations

It is not uncommon for variation to exist within individuals over short periods of time. As discussed
in the Chapter 3, the spinal discs thin over the course of the day, meaning that individuals are taller
in the morning. Significant changes in diet, state of health or exercise routines can also contribute to
intra-variations over short time periods.

Gender

Between (biological) genders, significant variation in body sizes can be identified. Typically,
females have lower measurement values than men across the gender-separated spectrum of most
body measurements (Figure 4.9), with the width of the hips being an exception. Another obvious variation between genders is the difference in body anatomy, which sometimes requires
separate standardization principles for how to measure specific (usually static) dimensions.
Variations also exist in the degree of muscularity, level of oxygen consumption and the location
and quantity of body fat. Given the increasing number of women in the industrialized workforce today compared with the past few decades, it is important that workplaces are designed to
suit the characteristics of both men and women. Figure 4.10 highlights the variation in stature
between genders.

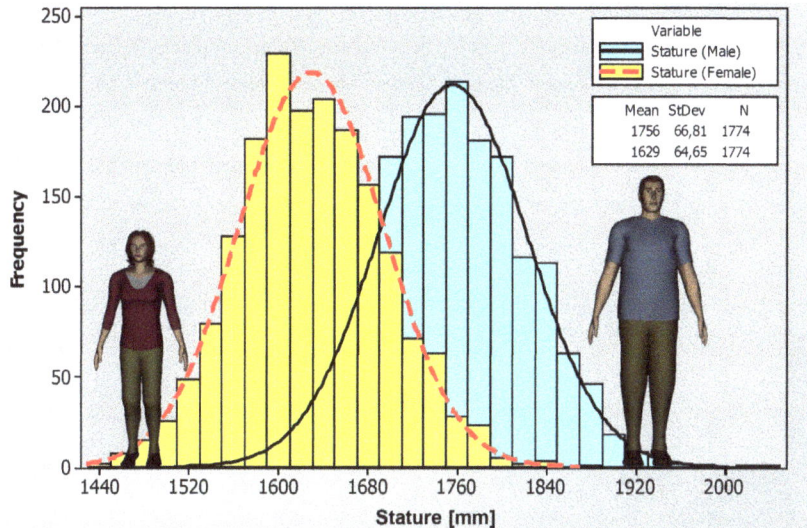

Figure 4.10: Variation in stature between genders (Brolin, 2013).

Image reproduced with permission from: E. Brolin. All rights reserved.

Nationality

Differences in nationality also contribute to variation between data sets. For instance a piece of equipment designed to fit 90% of the male US population would roughly fit 90% of Germans, 80% of Frenchmen, 65% of Italians, 45% of Japanese, 25% of Thais and 10% of Vietnamese workers. Given the increasing rate of diversity in the workplace, it is important to ensure that people from a number of geographic locations can work together in a healthy and safe environment. When designing for a European population, it has been common practice to take Dutch males and Italian females as the two extremes of the size spectrum.

Age

Age is another factor that plays a significant role in the variation between populations and measurements. Humans tend to be at their physical peak between 20–25 years old; at around the age of 30, some deterioration starts to occur, which becomes more prevalent in the later years (65 plus), as shown in Figure 4.10. These deteriorations typically mean: lower muscular and skeletal strength, reduced oxygen consumption, poorer eyesight and hearing, and increased sensitivity to vibrations, heat and cold. With increasing age, changes in stature have also been observed with spinal disc compression over time, leading to decreased height (see Figure 7.8). Given the increasing aging population in the workforce, it is important that these factors are considered in the design of workstations to maximise performance and minimise injury risks.

In addition to differences between various populations at a fixed time, it is also interesting to note changes in measurements over time. These days it's common for children to grow to be taller than their parents; this is in line with a recognized trend that people today are typically taller than their ancestors. An increase in stature of 10 mm per decade in Europe and North America during the 20th century has been observed. Increased weight over time in certain populations has also been identified over the past century.

Given the high degree of variation between populations, it is not possible to design workstations that will be suited to the entire population, so it is generally accepted to disregard the extreme ends of the spectrum and design for 5th–95th percentile. Given that muscle and skeletal strength varies due to age, gender and health status, it is necessary to design workplaces and tools where muscular strength exertion is optimized, so the most efficiency can be achieved at the lowest level of effort.

4.9. Methods for measuring body dimensions

To ensure accuracy across data sets and avoid poor data management and unreliable data, measurements are collected by professional physical anthropologists following standardized procedures where possible. Historically data has been collected manually using a combination of tools including rulers, anthropometers (a device used for measuring body segments), goniometers and calipers (Figure 4.12). However, with recent developments in technology full-body laser scanning, this is becoming increasingly popular, enabling all surfaces of the human body to be captured quickly in three dimensions. Measurements are taken in a number of predetermined postures, without shoes on and with as few items of clothing on as possible, to gain as accurate a representation as possible.

4.10. Anthropometric datasets

Extensive work has been carried out to measure different populations and obtain complete data sets that can be statistically analysed then scaled to provide an accurate representation of an entire

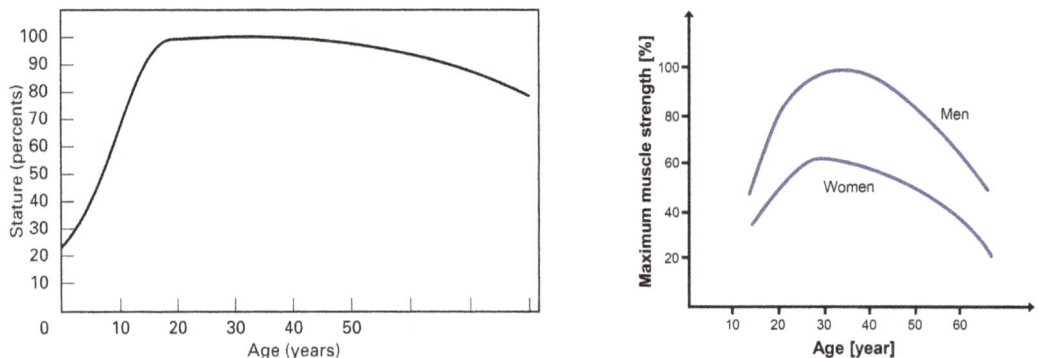

Figure 4.11: Variation in stature and muscle strength over time as a human ages (figures from Brolin, 2013).

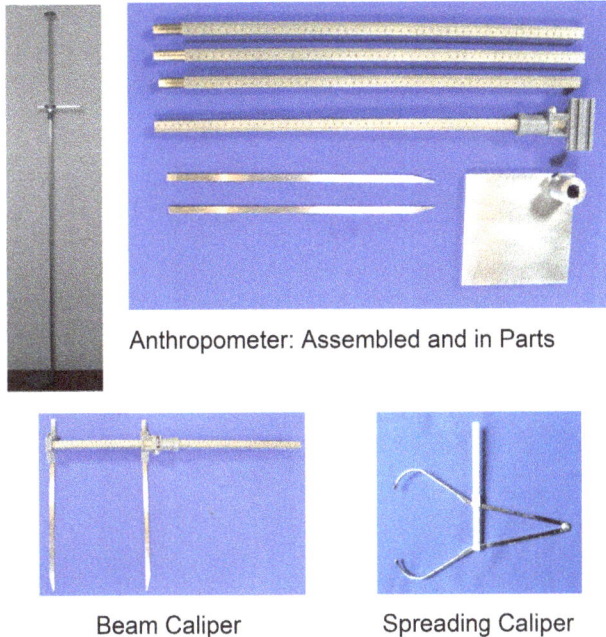

Anthropometer: Assembled and in Parts

Beam Caliper   Spreading Caliper

Figure 4.12: Examples of equipment for anthropometric measurements: anthropometer (top) and calipers (bottom). Figures from Gordon et al. 2014 p. 11.

Image permission granted by U.S. Army Natick Research, Development, & Engineering Center. The images have been slightly modified from the original (figure labels removed and some cropping).

population. However, obtaining large amounts of accurate data and recruiting people for body measurement is no easy task, and with emerging population trends, data can become obsolete. Datasets representing civilian populations are quite limited, but there is substantial data available from the US military. Historically the military has always had a wealth of available data, as numerous measurements were systematically taken by paid and qualified medical personnel for uniforms, weapons, vehicles and other equipment. However, given that the majority of soldiers are young, fit, healthy and historically male, it is difficult to generalise such data and create an accurate picture for the rest of the population. While hand, head and foot measurements are reportedly similar for both soldiers and civilians alike, other data shows little similarity. Some databases containing datasets for various populations are available; *Bodyspace* by Pheasant & Haslegrave (2006) is one of the most popular textbooks in this field, containing a number of measurements from anthropometric surveys.

Various online databases also exist, for example:

- openerg.com/psz/
- antropometri.se
- dined.io.tudelft.nl/dined/
- openlab.psu.edu

4.11. Design principles

Usually, it is not feasible to design workplaces to suit everyone perfectly from the shortest to the tallest, so a decision needs to be made about which members of the population will be eliminated. A commonly accepted rule is that the extreme sizes are eliminated and designs are based on measurements from 5th percentile females up to 95th percentile males; however, as we have already seen in the case of multivariate design, this can mean more than 10% of the population is excluded, so it is not sufficient to apply one standard rule; rather, it depends on the specifics of each design case. When designing workplaces they should be suited to both male and female Europeans aged between 18 and 70. In reality it isn't sufficient to only take into account anthropometric data; one must also consider behavioural patterns of people in different environments. This is why observations and participatory ergonomics are key sources of input during redesigns.

There are certain principles that can always be applied when designing for specific situations, which will be discussed in more detail below:

• Designing for the extremes
• Designing for adjustability
• Designing work heights

4.12. Designing for the extremes

When designing workspaces it's important to ensure there is enough space for employees to move around, especially given the varying nature of assembly tasks. So in this case the design should be based on values for the 95th percentile male so that there is sufficient space to accommodate their arms and legs and clearance above their head level so they aren't constantly hunched over. At the other end of the scale, where the issues are reaching components on the work surface and the strength needed to carry out the tasks, datasets corresponding to 5th percentile females should be used. Theoretically, adopting such a design philosophy should accommodate workers with body measurements closer to the median; however, testing and simulations should be done to confirm this before implementing the workplace. Guidelines exist to aid in the design of workplaces, e.g. the AFS (*Arbetarskyddsstyrelsens Författningssamling*) guidelines from the Swedish Work Environment Authority (Figure 4.13).

4.13. Designing for adjustability

In some instances it is not possible to accommodate everyone across the size spectrum; in such circumstances, adjustable equipment with varying height ranges should be added to the workstation. Where adjustable workstations are impractical, non-slip platforms are another possible addition to enable a more diverse workforce to work at the same workstation.

4.14. Designing work heights

One of the key areas that you have to consider when designing workstation layouts is the working height. Given the high degree of standing work on the production line, this is an attribute that affects

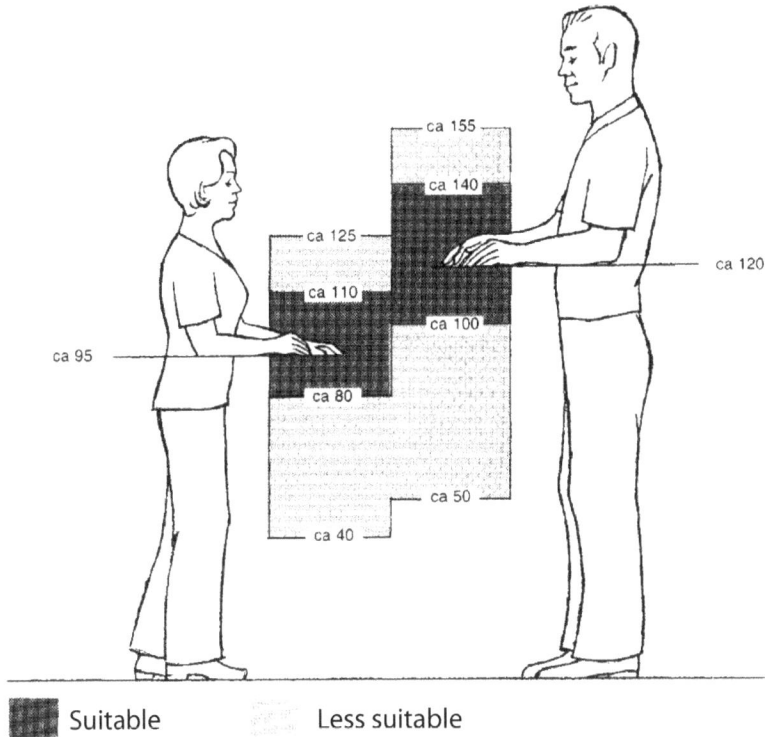

Figure 4.13: Workplace height design guidelines showing that the overlap between the tallest and short-est workers' ideal work heights may be rather slim (Swedish Work Environment Authority, 1998).

Image reproduced with permission from: the Swedish Work Environment Authority. All rights reserved.

all members of the workforce, so care should be taken to get it right and eliminate any injury risks. As discussed in Chapter 2, the shoulder is a complex structure prone to injury, so having workplaces set too high will force workers to continuously lift up their shoulders or work for prolonged periods with their arms extended above shoulder level. However, if the workstation is too low, the worker will be bent forwards and loading their back, which can also be an injury trigger. Adopting a body position like the one shown in Figure 4.13, with the arm bent at a right angle at the elbow, is regarded as the best option for light work. If a higher degree of precision is necessary, then the working height should be slightly higher, enabling the worker to see exactly what they are doing without straining their neck. For heavier work involving physical exertion the working height should be lower. Given that the component being assembled has its own height and that fixtures are often used to hold it in place, the workbench should be set at a height that takes this into account – meaning that while on its own the workbench might appear to be too low, but in reality while the worker is carrying out their assembly tasks it will be appropriate. Given that too high a workstation could lead to shoulder inju-ries, while too low a working height could result in back injuries, it is crucial to select the appropriate measurements to maintain an efficient and healthy workplace. An AFS guideline from the Swedish Work Environment Authority exists to aid in the design of workstation reaches and heights as shown in Figure 4.4 and Figure 4.13.

Steps for using anthropometric data in workstation design

1. Identify the necessary body dimensions needed for each element of the workspace design. For instance, hand length affects handle size, and eye height is relevant for information displays.
2. Identify the specific population of interest (age, gender, nationality) and determine suitable percentile ranges for each measurement.
3. Find a suitable anthropometric database with relevant measurements, if one is not available you may have to extrapolate data from another dataset or collect your own measurements.
4. Make a model of the proposed design based on the selected data; both physical models and computer simulations can be used to test the design.
5. Evaluate whether one fixed design will be sufficient, or if adjustable equipment needs to be added to accommodate the whole working population.

Study questions

Warm-up:

Q4.1)   Why would you choose to base measurements on a particular anthropometric database, such as one of the ones listed in section 4.4? Give at least two reasons.

Q4.2)   Explain what it means to "design for the 5th to 95th percentile" of a population.

Q4.3)   Explain the difference between static and dynamic body measurements.

Q4.4)   Name two examples of normally distributed body measures.

Q4.5)   Name two examples of non-normally distributed body measures.

Q4.6)   Why would it be a bad idea to design a workplace based on a fictive person with "average" measurements?

Q4.7)   What is a "critical user"?

Look around you:

Q4.8)   See if you can find examples of certain elements in a work environment that are not designed for a range of body sizes.

Q4.9)   Go into a kitchen – can you list elements of the environment that appear to have been designed with a particular body size in mind? Can you think of reasons why those measurements were decided upon? Who would have difficulty using the kitchen?

Connect this knowledge to an improvement project

- Think about the range of users who will use the workplace you are designing. Who are the tallest and shortest? The strongest and weakest? The most and least mobile (in terms of movement?). List the "extremes" for each task.
- List the critical tasks and the demands they place on human (or machine) performance – what are the maximum and minimum strength requirements? Reach distances? Hand clearances?
- Decide on whether you should design the workplace to offer adaptability (being able to change reach distances, work heights, choosing different size hand tools, etc.) or to design for a "critical user" for whom the work becomes impossible if the dimensions are not adapted to them.

Connection to other topics in this book:

- Knowing what tasks are typical in this workplace (Chapter 7) will help you figure out the requirements for strength, reach, manoeuvring space, clearance for hands, etc.
- Some ergonomics evaluation methods are only guaranteed to be valid for a certain population – for example, the NIOSH lifting equation is based on strength limits that have been measured mainly for males, so the method is only said to be 75% valid for females. For more, see Chapter 8.

Summary

- Work environments should be designed according to the characteristics of the human body, based on anthropometric data.
- Body measurements are described in terms of percentiles and what percentage of a defined population has which measurements.
- Variation between measurements is due to: poor data management, inter- individual variation, gender, nationality and age.
- Databases containing a wealth of measurement data collected using manually methods or body scanning, exist to aid designers.
- There is no such thing as the "average person" in all respects, so it is not a good idea to design workplaces based on their measurements.
- Design to exclude as few people as possible.

4.15. References

Brolin, E. (2012). Consideration of anthropometric diversity. Licentiate thesis, Chalmers University of Technology, Sweden.

Brolin, E. (2013). Anthropometry. [Lecture] Chalmers University of Technology, 18th February 2013.

Eastman Kodak Company (Ed.). (1983). *Ergonomic Design for People at Work*. New York, NY: Van Nostrand Reinhold.

Gordon, C. C., Blackwell, C. L., Bradtmiller, B., Parham, J. L., Barrientos, P., Paquette, S. P., Corner, B. D., Carson, J. M., Venezia, J. C., Rockwell, B. M., Mucher, M. & Kristensen, S. (2014). *2012 Anthropometric Survey of US Army Personnel: Methods and Summary Statistics* (No. NATICK/TR-15/007). Army Natick Soldier Research Development And Engineering Center, MA.

Hanson, L., Sperling, L., Gard, G., Ipsen, S., & Vergara, C. O. (2009). Swedish anthropometrics for product and workplace design. *Applied ergonomics*, 40(4): 797–806.

Kroemer, K. H., Kroemer, H. J., &. Kroemer-Elbert, K. E. (2010). *Engineering Physiology*, 4th edition. Berlin: Springer-Verlag.

Kroemer, K. H. E. & Grandjean, E. (1997). *Fitting the Task to the Human: A Textbook of Occupational Ergonomics*. London; Bristol, PA: Taylor & Francis.

Pheasant, S. & Haslegrave, C. M. (2006). *Bodyspace: Anthropometry, Ergonomics and the Design of Work*. 3rd edition. Boca Raton, FL: CRC Press.

Singleton, W. T. (1964). A Preliminary Study of a Capstan Lathe. *International Journal of Production Research, 3*(3): 213.

Swedish Work Environment Authority. (1998). Ergonomics for the Prevention of Musculoskeletal Disorders – Provisions of the Swedish National Board of Occupational Safety and Health on Ergonomics for the Prevention of Musculoskeletal Disorders, together with the Board's General Recommendations on the Implementation of the Provisions. Solna, Sweden: Arbetarskyddsstyrelsen. Legal provision.

Bibliography

Bohgard, M. (Ed.) (2009). *Work and Technology on Human Terms*. Stockholm: Prevent. ISBN 978-91-7365-058-8

Heinz, G., Peterson, L. J., Johnson, R. W. & Kerk, C. J. (2003). Exploring relationships in body dimensions. *Journal of Statistics Education* 11(2).

CHAPTER 5

Cognitive Ergonomics

THIS CHAPTER PROVIDES:

- Descriptions of how the human brain and our perceptive abilities work and respond to stimuli and mental workload.
- Some guidelines for good cognitive design of instructions, interfaces and cognitive assembly supports like fixtures.

How to cite this book chapter:
Berlin, C and Adams C 2017 *Production Ergonomics: Designing Work Systems to Support Optimal Human Performance.* Pp. 83–106. London: Ubiquity Press. DOI: https://doi.org/10.5334/bbe.e. License: CC-BY 4.0

WHY DO I NEED TO KNOW THIS AS AN ENGINEER?

Cognitive aspects of a workplace concern the sensory signals that give our brains the clues and cues to understand a task or to solve a problem. Your task as an engineer is to create the best possible conditions for workers to correctly interpret the task and task status, in order to avoid danger, errors, confusion, irritation and mental overload. This is obviously a very powerful design area, which can make or break a worker's ability to understand what to do in the workplace. In other words, we are moving focus from the physical to the mental in this chapter. Many cognitive aspects have to do with our interpretation of sensory stimuli (vision, hearing, touch, smell and taste), our capability to recognize patterns, our understanding of instructions and our ability to associate symbols with meaning. The brain is constantly handling cognitive processes (even during sleep!) and often needs to be well-rested and nourished to work optimally. However, it is not uncommon that work is performed in a state of fatigue, which adds limitations to our cognition, attention, perception, memory and mental models.

With some basic knowledge of good cognitive design principles, a production engineer can minimise unnecessary mental workload and help an operator perform their work tasks more efficiently and with fewer errors and misinterpretations. This theoretical knowledge can contribute to the design of workplaces, instructions, machines, tools and activities that communicate better to the worker how to achieve their goals. This chapter also brings up some examples of currently existing cognitive support solutions used in modern production.

WHICH ROLES BENEFIT FROM THIS KNOWLEDGE?

The *system performance improver* who understands human cognitive abilities and limitations will be able to specify requirements for appropriate equipment, instructions and human-machine interfaces that can aid workers in doing tasks efficiently and correctly.

The *purchaser* will be able to better understand the value of investing in human-machine systems that transmit information and instructions as quickly and intuitively as possible. For both these roles, there is an economic argument that workers with good cognitive support commit fewer errors, leading to better product quality and less waste and scrap – but this may need to be proven with a business case and translated into a prospect of higher quality and/or productivity to convince a purchaser.

The *work environment / safety specialist* can use this knowledge to pinpoint safety hazards and risks for error that can be traced to signals and information being missed or misinterpreted, due to sensory distraction or insufficient cognitive support.

5.1. What cognitive limitations exist in the workplace?

Human workers are still preferred in many assemblies over robots because of their superior ability to respond to variations in assembly instructions and quickly take decisions to address deviations from the normal process flow. However, the fact that the human is a thinking, learning, processing being that is constantly changing, also poses some consistency problems for performance. Sometimes, even on the basis of plenty of experience, humans can misinterpret information, make mistakes or make ill-advised choices, like deciding to take shortcuts in a process, which has in the past resulted in dire consequences such as costly, unnecessary mistakes, or even fatal consequences for health and safety.

One extreme example is the partial nuclear meltdown of the Three Mile Island power plant in Pennsylvania, USA, in 1979. (United States Nuclear Regulatory Commission, 2013). This emergency is attributed to operator error and several human factors errors that caused the plant operators to misunderstand the process, ignore status alerts, miss alarm signals and shut down the wrong functions. The bad cognitive ergonomics of the plant schematics (instructions), machine interfaces and alarm signal system led to partial core meltdown, radioactive contamination and enormous public distrust and backlash against the nuclear energy sector. Following this and similar accidents in other countries, the nuclear sector has globally invested large and lasting efforts in improving human factors aspects of its technology, knowledge among its personnel, and tightly controlled safety aspects (United States Nuclear Regulatory Commission, 2013). It is today one of the most advanced industrial sectors regarding human factors, with an emphasis on cognitive ergonomics.

In more production-related cases, the same assembly line and operators may be used to produce multiple variants of a product, where the fundamental elements are the same but subtle differences exist. This can often cause confusion or errors, leading operators to assemble parts incorrectly. This in turn causes defects and quality issues further down the line, resulting in unnecessary costs and rework. Many have attempted to address this issue with varying degrees of success through the use of various different methods, which will be outlined later in this chapter.

5.2. Human capabilities and limitations

Up to this point, this book has mainly focused on the human locomotive system and ways to improve physical well-being and performance. In this chapter, we focus on the abilities and limitations of the human mind and senses, which work together to process and interpret information from our environment and formulate goals for action – this is what constitutes a human's cognitive abilities (Figure 5.1).

Our mental capacity changes with age (both improving and declining, depending on training and genetic factors), and our cognitive abilities are a combination of skills, experience, pattern recognition, attention, memory, ability to focus, expectations, associations, generalization and the ability to sort information into categories. Of course, our physical well-being can have a significant impact on these abilities. If we try to perform mentally intensive work tasks when we are tired, over-stimulated, stressed, emotionally or chemically affected, alarmed, distressed or hungry, our brain may transfer from a mode of high-functioning thought (planning, reasoning, evaluating) to survival mode[1] (instinctive, quick actions to evade danger or discomfort), which may at worst result in negative effects ranging from small mistakes to fatal accidents. In particular, human abilities are drastically limited by being in a state of fatigue. Fatigue can contribute to mistakes and accidents, especially for tasks requiring sustained vigilance (such as observing a monotonous process that may change suddenly). This works both ways – poorly designed cognitive supports and tasks that routinely cause mental overload can also contribute to chronic fatigue, leading to demotivation, ill health and absenteeism.

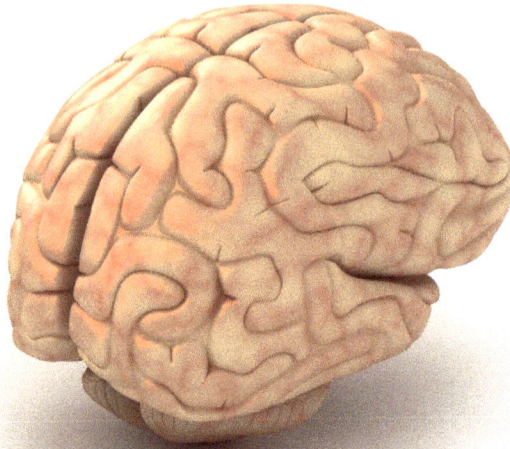

Figure 5.1: The human brain interprets information from the external environment.

A cognitively well-designed work system can lessen the impact of fatigue by minimizing the ability to perform incorrect or dangerous actions. The more demanding the task or the more stressful the situation, the more important carefully designed cues become.

5.3. The senses

Commonly, it is said that humans have five senses, which are (listed from most to least dominant): vision, hearing, touch, smell and taste. The senses convey information about our surroundings and the internal state of our own bodies, by receiving stimuli through different receptors and sending them via the nerves to the brain for processing. While the latter two senses are seldom intentionally used for communicating information (and are therefore not described here in depth), a combination of visual, auditory and tactile cues (e.g. vibrations) make up the majority of signals that are trans-mitted to humans in a workplace setting. Additionally, humans are said to have a sense of balance and muscle sense[2], both of which can be used to interpret our surroundings and act accordingly (for example the sensation of gravity telling us which way is up – however, this sense can be confused by contradicting signals from our other dominating senses).

Vision

Vision is the most dominant sense that humans use. Our field of vision in total extends about 170 degrees horizontally; the outer rim (peripheral) of that field is good at detecting movement but not detailed information. Therefore, we are dependent on viewing detailed information in our central field of vision.

Light is a form of electromagnetic radiation (see Chapter 12), whose different wavelengths are interpreted by human eyes as different colours. The visual photoreceptors in the eyes are called the *rods* and *cones*. The cones are very sensitive to small differences in shape and colour, but require good lighting to function, while the rods are much more numerous and more sensitive to seeing in dim light, but they cannot distinguish colour. Our sense of vision is connected with our perception, which is actively looking for patterns and structures that our mental bank can recognize as meaningful. Several parameters affect sensory processing of light (Table 5.1).

As we age, our visual abilities tend to deteriorate from the approximate age of 40, especially our capacity to detect low contrast, small symbols and weak stimuli. This makes good task lighting and clear visual cues (with sufficient size and time duration) extra important when designing tasks, interfaces and environments for a whole work population (Bohgard, 2009 p.p. 346–350). Table 5.2 includes some design principles for visual information.

Table 5.1: Parameters that influence vision (adapted from Bohgard, 2009 pp. 351–352).

CONTRAST	Contrast sensitivity is our ability to distinguish between light and dark, allowing us to see lines, text, shapes and contours of objects. High contrast means that there is a large difference between black and white in the field of view, while low contrast means more subtle differences on a grey scale. Our contrast sensitivity decreases as we age, meaning that it becomes more important that readable information should have enough differentiation between black and white.
COLOUR	Colour is the result of how our brains interpret and distinguish different wavelengths of light. The receptors that dominate colour vision are the cones, which are centrally located in the eye's retina. Different people have different abilities to distinguish and interpret different colours, depending on age, education, culture and genetic preconditions (such as colour blindness).
DARK-ADAPTED VISION	Depending on the number of rods (the more numerous and sensitive receptors) in the eyes, our ability to see in the dark varies, with some people experiencing a brief period of inability to see. Also, in dark environments, our ability to see different colours decreases.
DEPTH PERCEPTION	This is the ability to distinguish how far away different objects are relative to each other. Our ability to perceive depth is dependent on binocular (two-eyed) stereo vision and on previous experience and is decreased in the dark.
MOVEMENT DETECTION	The human eye is very good at detecting movement (a remnant of our descent from stone-age hunter-gatherers), which can sometimes be used as a deliberate way to attract attention to details or changes of status in a process. This is extensively used in software, for example progress bars and flashing advertisements on web pages.
GLARE	Glare is irrelevant high-intensity light that does not contribute to better illumination, but instead irritates and overwhelms our sense of vision, leading to temporary inability to see.

Table 5.2: Key design factors for presenting visual information (adapted from Bohgard, 2009 p.p. 351–352).

INTENSITY	Particularly for displays and signs, the amount of light entering the human eye must not cause glare, nor must it be too dimly lit for the eye to perceive contrast and colour.
CHOICE OF COLOUR	It is wise to be restrictive with colour-coding critical information, or to provide a redundant backup system for interpreting the colours correctly. For example, colour-blind people who cannot distinguish red and green colours close together can still interpret traffic lights correctly because the colours are separated and follow a consistent rule of where they are positioned.
STRENGTH OF LIGHTING	Different tasks require different lighting strengths to be sufficient. For example, high-precision detail work demands much higher light compared to general office work lighting. Recommended lighting levels for different types of work are available (more on this in Chapter 12).
CONTRAST	Sufficient contrast – i.e. difference in object luminance – is important for humans to be able to distinguish symbols from their background, especially regarding written information and alarm signals.
ANGLE OF VISION	Consider where in the human's field of vision information must be placed to be perceived, the appropriate distance away from the eyes, and the angle that the neck must adopt to see well. This should be designed in parallel with illumination of the object being viewed.

Hearing

Human hearing, like vision, is tightly coupled to our cognitive pattern recognition skills, which helps us distinguish many nuances of sound – most of us can correctly identify the direction a sound comes from, the volume, the pitch (allowing recognition of melodies) and even when certain signals concern us or not, such as when hearing our name spoken in a noisy environment or being able to filter out sounds that carry no meaning for us (in some cases known as selective hearing). Sound is a particularly effective complement to vision when we are overloaded by visual stimuli. Sound can be used to bring attention to changes in process status, to warn of danger, to indicate distance (such as warning systems for backing a vehicle) or to confirm correct actions.

Since sound is a form of vibration, the body perceives audible sound via vibration of the inner ear, while non-audible sounds are perceived as vibrations. Particularly sub-sonic (low) frequencies have been known to cause whole-body vibrations that cause nausea and feelings of discomfort and depression. Also, it is important to remember that hearing abilities change with age – notably, there are high-pitched frequencies that can be heard primarily by young people, but this ability diminishes already in early adulthood. However, hearing loss can also occur as a result of exposure to noisy environments, but this is injury-driven hearing loss rather than age-related.

Table 5.3: Parameters that influence sound and hearing (adapted from Bohgard, 2009 p.p. 351–352).

LOUDNESS (AMPLITUDE)	Sound travels in waves, which have different amplitudes corresponding to loudness. The ear has limits for how much loudness it can tolerate before permanent hearing injuries occur.
PITCH (FREQUENCY)	Pitch or frequency (the wavelength of the sound) defines the "tone" of the sound, and differences in pitch delivered in a sequence can be distinguished by the human ear as melody or signals that can be associated with meaning. The human ear (and body) has sensitivity to a wide range of frequencies, but is unable to hear very high pitches well (such as dog-whistles).
LOCATION (DIRECTION)	Thanks to stereo hearing (involving both of our ears), humans can determine which direction a sound is coming from by interpreting the differences in loudness and pitch between the two ears. This ability is so exact that it is actually possible to create "sound illusions" that convince a listener that a sound source is moving in space. This is done by recording sound in a quiet room, using two separate microphones spaced apart by about the width of a human head.

Touch

The tactile sense, also known as *haptics*, is what allows us to perceive differences in pressure, temperature and frequency (as in vibrations) – most frequently through nerve receptors in our skin that are sensitive to stimulation from bending of hairs in the skin, and to pain. Particularly the hands are sensitive to very small sensations, but evolution has made most of our skin able to register slight touches (as light as that of a spider web).

5.4. Human cognitive processes

Cognition is the overall process of handling information. It is the combination of sensory stimulation, focus, perception, working memory, long-term memory and interpretation, leading to decision making and response.

There are two categories of mental processing of information; either the process is in response to sensory stimuli and is unconscious/automated (bottom-up), or it is a conscious chain based on desires, previous experience or knowledge, expectations and generalizations (top-down).

Attention

Attention means devoting a human's mental resources to a task or event at hand. Undivided attention focuses all our cognitive processing capability to one stimulus. When our attention is divided between two or more information sources, our ability to correctly process stimuli and interpret information is decreased.

Human attention functions best when events come at regular, relatively frequent intervals, but once activity frequency is too low, our attention levels fall, and there is a risk that small status changes or subtle signals will be missed (Bohgard, 2009). The ability to keep focus on a process for duration of time is called *alertness* or *vigilance*. Since humans are not naturally good at remaining vigilant for a long time, it is important to support attention using enough sensory stimulation, the right amount of pressure and the right frequency of activity. A lack of this support is called a *monotonous* task or environment and leads to boredom and decreased motivation. Boredom is a mental state where our brain deactivates certain nervous centres and the human experiences weariness, lethargy and decreased alertness (Kroemer and Grandjean, 1997 p. 219). In this state, humans are less ready to perform tasks well or respond to sudden stimuli. Lapses in attention can lead to quality losses, accidents and inferior performance.

Memory

Memory is the process that allows learning through storage of information, experiences and rules in the brain. It is divided into long-term memory and short-term/working memory (STM). Working memory allows us to store new information temporarily in order to make sense of patterns and relationships between data points and mentally process the information into coherent chunks that can be stored in the long-term memory. The short-term memory also allows us to recall recent events, up to a couple of hours ago. However, our short-term memory capacity is limited in how many new information points it can take in at once. An established rule of thumb is "The rule of 7", which states that 7 ± 2 is the maximum number of unrelated items the STM can store at the same time. It is possible to train the short-term memory performance to increase capacity, mainly by using a technique called "framing" which means actively identifying a pattern, category or sequence that groups or contextualizes the items into coherent chunks. Examples include associating items with a story, an experience or a theme.

One way to decrease the load on working STM (thereby) is to practice tasks and movements until they are stored in the long-term memory. This training decreases a human's sensitivity to stress by liberating working memory so that its limited capacity is no longer occupied by routine actions.

After information has been processed, the human brain has enormous storage capacities in its long-term memory. Recalling information from there can be either easy for strong memories or frequently practiced behaviours, but may sometimes be dependent on appropriate cues that stimulate recall of events and experiences months or years ago (Kroemer and Granjean, 1997 p. 180; Bohgard, 2009). When we perceive something, this often allows association to items in our long-term memory.

Memory can be categorized as in Table 5.4.

The ability to recall and store information both from short-term and long-term memory is deteriorated by stress, fatigue, hunger, disturbing sounds, etc. Particularly stress can affect the capacity of our STM to the point where tunnel vision occurs, leaving the human fixated on handling only one infor-

Table 5.4: Categorization of memory.

Declarative memory: requires active recall	Non-declarative memory: does not require active recall
• *Semantic memory* – meanings, concepts, understandings. *Examples:* language, abstract knowledge about the world • *Episodic memory* – past personal experiences and events, known as *autobiographical* memory. *Examples:* places, dates, times, associated emotions	• *Procedural memory* – also known as *implicit* memory. This type of memory is associated with motor learning and is not consciously recalled, but translates automatically to actions or movements. *Examples:* riding a bike or tying shoelaces. • *Perceptual memory* – Allows recognition of sensory stimuli as meaningful. *Examples:* recognizing faces, voices, smells.

mation source and unable to take in additional sensory stimuli, a situation which could prove to be dangerous. Some memory recall deterioration may also result from age, but this is highly individual – keeping the mind active and stimulated can lessen such effects.

Perception

Our capacity to take in information from the environment, associate it with meaning and mentally organize it is called perception. This capacity is based on previous recognition, knowledge and experiences, which gives us a basis for selecting, interpreting and categorizing information. This basis for making meaning is what creates our mental models, or expectations of how things appear. These preconceptions speed up our mental processing capacity, but also make us susceptible to illusions. Illusions are when our interpretation of sensory signals are mismatched with reality. Examples include optical illusions (Figure 5.2) and when the brain automatically filters out information that it has learned to sort as meaningless (Figure 5.3). It is important to note that our expectations and the context that information appears in greatly influence what the brain filters as meaningful information or categorizes as having a certain meaning.

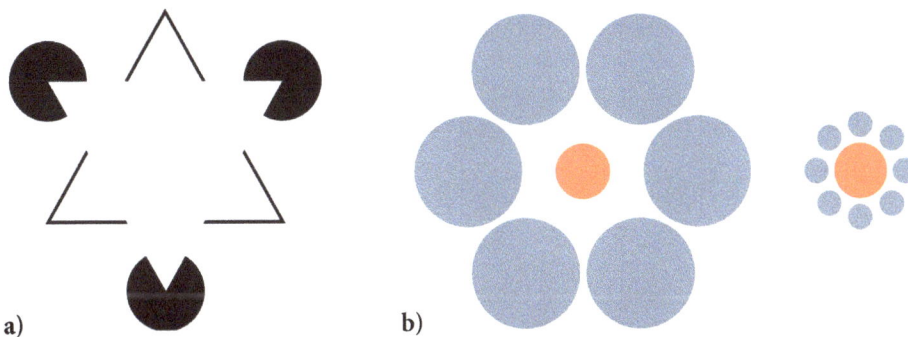

Figure 5.2: Examples of optical illusions: in some cases we a) see things that aren't necessarily there, as in the Kanisza triangle, or b) interpret the size of identical shapes (the orange dots are the same size) incorrectly due to confusing "clues" from surrounding information, as in the Ebbinghaus illusion.

Images sourced from Wikimedia Commons (both credited to Fibonacci/Wikimedia Commons, 2014a and 2014b).

Figure 5.3: Many people cannot see anything wrong with this book title. Can you?

Image by C. Berlin.

Mental models

The idea of "mental models" provides a language for speaking about expectations that people have. What is important is that these expectations sometimes lead a worker or user to look for specific cues in order to interpret their surroundings or a machine interface, and will then interpret them according to previous knowledge and experiences. There may sometimes be a mismatch of the mental model with the actual reality, which can result in errors or mistakes. Therefore it is very important to clearly convey the correct mental model of a tool, product or system. This is especially true for instructions and in education or training.

For example, the word "tree" is a visual or auditory signal that carries a symbolic meaning, but that symbol does not actually tell us what the tree looks like, how tall it is, how wide, how leafy, etc. Most human adults have learned that a tree can appear in many different ways and still be classified as a tree; this means that the mental model of a tree has some degree of uncertainty, but can generally be expected to have a trunk, branches, and leaves.

However, "Palm tree" may set different expectations on what that tree looks like, and depending on what kind of palm tree a human has encountered before, they may expect anything from a very long, smooth trunk with long leaves and coconuts at the top, to a short, wide mess of rough, fibrous trunk exploding into an array of fan-shaped pleated leaves. These are examples of two different mental models of a palm tree, and they are based on knowledge, experience and cultural background.

The same can be said of mental models of a "factory": in different minds, the expectations of what a factory looks like may be very different, depending on familiarity with different sectors, exposure to historical or cultural representations, childhood encounters, etc.

5.5. The role of expertise: The SRK model and types of mistakes

As we practice anything, be it physical or cognitive, we gradually develop skills. We progress from being a novice (a beginner) who is dependent on instruction, via an intermediate state where become less dependent on confirmation that we are doing things right, to a highly skilled state of being an expert, where actions, rules of thumb and cause-effect relationships have been internalized and stored in our long-term memory (or as some say figuratively, in our bones).

This progression from novice to expert has been described by a classic theory known as Rasmussen's SRK model, short for Skills-Rules-Knowledge (Rasmussen, 1983). This theory states that humans make decisions and solve problems in three different modes:

- **S** – Skill-based: actions performed without consciousness and routine actions; sensory- motoric responses
- **R** – Rule-based: actions governed by rules, procedures and old knowledge; involves recognizing signs and associating to related process status, which is then associated with stored rules.
- **K** – Knowledge-based: actions which require explicit thinking and problem solving; based on identifying the meaning of symbols and making a plan. Includes trial and error.

(Adapted from Rasmussen, 1983)

A novice frequently operates in the knowledge-based mode, dominated by cognitive processes that are highly dependent on short-term working memory, which is limited in how much new information it can store. Novices can therefore need more time to interpret and perform tasks, be more easily overwhelmed or mentally overloaded by their work environment and tempo, and make more mistakes based on forgetting or misinterpreting rules. As they learn, intermediate novices gradually adopt more rule-based action, which is based on accumulated, stored knowledge. An expert, on the other hand, acts and reacts almost instinctively in the skill-based mode, with greater task speed and fewer mistakes. When errors do occur, they are more like slips or lapses of concentration – simply put, "sloppiness".

A related theory by Reason (1990), characterizes different types of errors that relate to different levels of cognitive processing:

- **Slip**: correct plan but incorrect action; easily observable.
- **Lapse**: correct plan but incorrect action; more unnoticed causes (such as forgetting).
- **Mistake**: incorrect plan; caused by incomplete or incorrect knowledge.

(Adapted from Reason, 1990)

5.6. Mental workload

Measuring mental workload is complex because an assessment of total workload should, to be fair, take consideration of all the different components of cognition and of surrounding factors that are known to influence mental performance. Also, many of these factors are hard to measure objectively, which leads to the conclusion that measuring mental workload is often a measurement of the individual's perception of it. However, this information is still useful, as it can be used as a before-and-after type baseline for evaluating cognitive ergonomics improvements. If a measurement is made before a change and the individuals asked perceive an improvement after it, the mental workload can be interpreted as lessened for that workforce.

NASA (NASA, n.d. and Hart and Staveland, 1988) has developed a rough questionnaire method for measuring total mental and physical workload, called NASA-TLX (Figure 5.4). This lets individuals rate their workplace or task with regards to six different components of physical and cognitive loading and support.

5.7. Designing to support human mental capabilities

Design principles

The following sections offer specific design principles geared at supporting the human cognitive capabilities of attention, perception, memory and mental models. This list is adapted from a more extensive one by Bohgard (2009 pp. 394–399).

The thirteen design principles are:

1. Minimize time and effort for finding information
2. Proximity/closeness
3. Engage multiple senses
4. Legible displays
5. Appropriate number of information levels
6. Avoid only knowledge-based data
7. Redundancy
8. Avoid similar objects

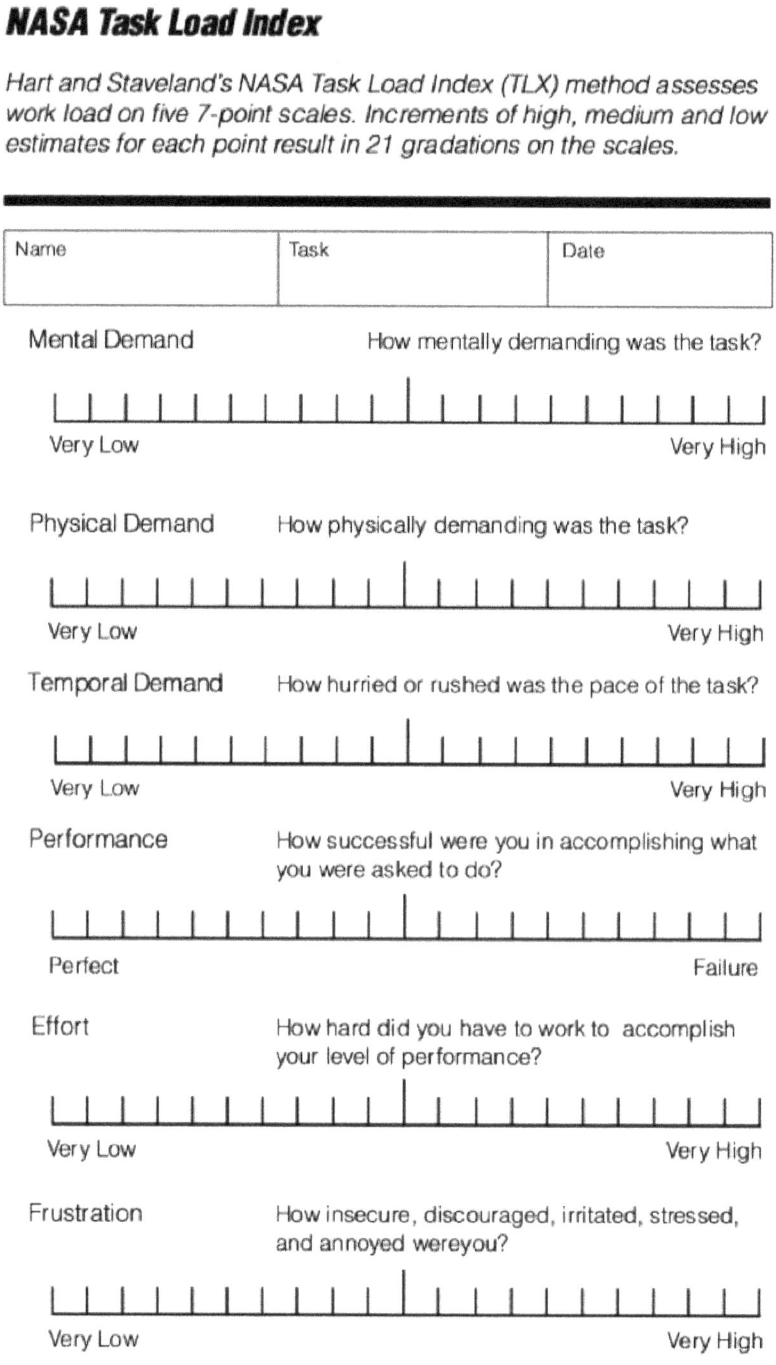

NASA Task Load Index

Hart and Staveland's NASA Task Load Index (TLX) method assesses work load on five 7-point scales. Increments of high, medium and low estimates for each point result in 21 gradations on the scales.

Name	Task	Date

Mental Demand   How mentally demanding was the task?

Very Low   Very High

Physical Demand   How physically demanding was the task?

Very Low   Very High

Temporal Demand   How hurried or rushed was the pace of the task?

Very Low   Very High

Performance   How successful were you in accomplishing what you were asked to do?

Perfect   Failure

Effort   How hard did you have to work to accomplish your level of performance?

Very Low   Very High

Frustration   How insecure, discouraged, irritated, stressed, and annoyed wereyou?

Very Low   Very High

Figure 5.4: The NASA-TLX form with its six sub-scales (NASA, 2014).

Source: NASA Ames Research Center; Used with permission.

9. Minimize the amount of short-term memory data
10. Show anticipated system status
11. Consistent/natural representation
12. Illustrated realism
13. Show movable objects for dynamic information

Table 5.5: Key design principles for supporting attention (adapted from Bohgard, 2009).

1. Minimize time and effort for finding information	Efficiency and motivation of work decreases when too much time and effort is spent searching for relevant information. This includes having to look for information in different places, in different menus, displays, etc. Frequently used information should be easily accessible and emphasized, and thematically related information should be grouped together.
2. Proximity/closeness	Similar or related information sources should be visually linked. Use physical nearness; indicators such as lines, arrows or boxes; or a uniform format of colour, pattern, shape, typeface or the like. In the case of auditory signals, use easily recognizable differences in pitch, loudness, repetition rate, rhythm and melody to distinguish similar and non-similar information.
3. Engage multiple senses	When needing to pay attention to large amounts of simultaneous information, it helps to engage multiple senses. Alternate between vision, sound and touch to deliver different types of signals.

Supporting perception

Table 5.6: Key design principles for supporting perception (adapted from Bohgard, 2009).

4. Legible displays	Legibility means "possibility to read". Support perception of text using high contrast, appropriate illumination, sufficiently large text, a clear font and the correct viewing angle.
	From an auditory perspective, use clearly distinguishable sounds (for example use a clearly different pitch to the surrounding ambient sounds that normally occur in the environment) – as for loudness, remember that the purpose is to convey information, not to jolt, scare or distract the listener. For both of these aspects, be careful to design so that legibility is possible for elderly workers with decreasing vision and hearing.
5. Appropriate number of information levels	It is advisable to limit levels of information to three, since increasingly nested structures challenge expectations of where to look for information and take a long time to search through. To ensure that colour-vision impaired can distinguish between colours in an interface, it is advisable to limit colour codes to two. It is also hard to distinguish more than five levels of line thickness, shape differences or fonts. (Sounds cannot, as a rule, be presented hierarchically, since they are transient in time.)

6. Avoid only knowledge-based data	Quite frequently, the preconceptions and expectations of a human will override purely responsive reactions to stimuli; therefore, to ensure that unexpected signals are correctly interpreted, such messages must be reinforced and emphasized, for example with more central placement on a display, flashing, size increase or colour change.
7. Redundancy	It is possible to reinforce accurate interpretation of a message if it is presented in more than one way, using several modalities or senses (for example, both visual and auditory cues can be used for alarms) or more than one sensory representation in the same sensory domain (e.g. image + text, shape + colour or sound signal + voice).
8. Avoid similar objects	When stimuli (such as objects, symbols or sounds) appear to be similar, the brain associates the same meaning to them, which may lead to confusion or misinterpretation if they have different functions or meanings. Therefore, it is important to signal differences in function with clear differences in appearance, size, duration, placement, structure, etc.

Supporting memory

Table 5.7: Key design principles for supporting memory (adapted from Bohgard, 2009).

9. Minimize the amount of short-term memory data	As far as possible, free the short-term memory from loading. Use the operators' "real-world knowledge" and learned behaviours to lessen dependency on working memory resources. Use the idea of the "magical number" 7 ± 2 as a maximum for simultaneous sensory stimuli.
10. Show anticipated system status	Design the system or interface to signal future states (for example showing a progress bar in software). This removes the mental load from the operator of calculating or guessing what will happen next based on available data, and makes the task into a simpler perceptive one. Letting the system do the forecasting frees the operator's mental capacity and supports *proactive* action; in the opposite case, a mentally overloaded operator can only respond *reactively*. For sounds, designed transitions in loudness or pitch can indicate changes in state over time (for example the decreasing pitch of a running-down motor).
11. Consistent/natural representation	If operators (usually experienced ones with many learned routine behaviours) are used to a particular configuration or interface design, changing the design too drastically from the familiar layout (such as changing colour coding) may be a source of mistakes and slips. New designs should correspond to learned rules and interpretations among the operators.

Supporting mental models

Table 5.8: Key design principles for supporting mental models (adapted from Bohgard, 2009).

12. Illustrated realism	Use visual cues that correspond to reality when designing information – the aim is to correspond to the operator's mental model of a measurement (such as showing temperature on a vertical scale) or a place (such as on a process status display, where machine statuses should be arranged the same way as the machines in reality).
13. Show movable objects for dynamic information	Use animation, sound modulation and other dynamic representations, making sure that movement indicating status changes over time match the operator's mental model of the process change. For example, a sound that decreases in pitch might correspond to a sinking or lowering movement.

5.8. Cognitive ergonomics supports used in industrial production

Having introduced the concept of cognitive ergonomics and the capabilities of the human mind, we will now bring it closer to home and look at how this topic affects the operator in production industry. Many different tools and methods that aid the operator from a cognitive perspective exist in the assembly environment, limiting the mental capacity required. Interestingly, a number of these methods came about purely from the desire to optimize the performance of systems, rather than to specifically provide operators with cognitive support; the added cognitive benefits sort of came about as an added bonus almost unintentionally. This section will introduce various different ways in which cognitive ergonomic considerations are effectively being used in the production environment. The key aspects we will discuss are:

- Design for assembly
- The use of fixtures
- Kitting
- Standardized work
- Work instructions
- Poka yoke
- Pick by barcodes
- Pick by light
- Pick by voice
- Andon systems

It is important to note that no single solution is accepted as the go-to standard approach; rather, the solution is dependent on the nature of each individual business at that time, given their unique requirements, size, strengths and weaknesses. In some cases, businesses choose to adopt one approach at one time and then switch to another or multiple approaches as new issues arise within their business activities. Typically, this decision depends on characteristics such as cost, quality, delivery time and delivery time reliability, production system flexibility, and product flexibility. Certain concerns are generally considered to have a higher priority than others at different times. All of these support systems are based on the idea that it should be hard to do things wrong.

5.9. Design for Assembly

A recurring problem in industry is that all too often the product is designed without consideration of the fact that the product has to be put together by an assembler in a production facility. Design for assembly (DFA) is a method which aims to encourage designers to think about the assembly implications of their design, for instance by minimizing the number of required components and enabling as simple an assembly method as possible (Boothroyd, 2002). This should in turn lead to reduced times and cost during the manufacturing stage, while maintaining quality.

DFA aims to enhance the level of communication between the manufacturing and design teams, to ensure an optimized solution meeting the requirements of both parties is achieved. Taking DFA into consideration during all stages of the product's design and development right from its conception reduces the need to make design changes late in the process. The DFA procedures and design rules that should be followed differ depending on whether the product is manually or automatically assembled. The general DFA guidelines try to address two key areas, the handling of parts and the way in which parts are connected or fastened.

The basic concepts of the DFA methodology will be briefly introduced below; however, to gain a full understanding of this method, the exact details on how to carry it out on a real product and the quantitative tools that exist can be found in Boothroyd's *Product Design for Manufacture and Assembly* (2002).

The following are general guidelines that should be considered during the design of products, as they will have a positive impact on the assembly stage of the product, aiding the operator with their work tasks from both a physical and cognitive perspective.

Where possible, parts should:

• Use geometrical features: symmetrical, or obviously asymmetric for instances where symmetry can't be achieved.
• Design parts that cannot be attached incorrectly.
• Use shapes or features that ensure parts won't stick together when in mass storage containers.
• Avoid shapes or features that will cause parts to tangle when in mass storage containers.
• Easy to handle, avoid very small or excessively large, slippery or sharp parts that could be difficult or hazardous to handle.
• Reduce the count and part types.
• Ensure sufficient access and visibility is provided

Having symmetrical or obviously asymmetrical shaped parts will ease the task of the assembler and reduce mental load as the way in which the product should be assembled is much more obvious, so to some extent the shape of the part acts as an unspoken intuitive work instruction to the assembler. It also contributes to time saving in assembly as it reduces or eliminates the need for the operator to reorient parts during assembly. The other guidelines are more related to physical ergonomics considerations and the reduction of poor postures and potential frustration areas for the operator, such as constantly having to spend time untangling small springs from each other or straining their neck to ensure parts are aligned and attached correctly.

5.10. The use of fixtures

Providing assemblers with nothing but a table and a few tools would likely result in high levels of frustration, dissatisfaction, disorder, confusion, poor posture, MSDs and eventually absenteeism. To remedy these problems, carefully designed fixtures are installed at workstations to ease the mental workload on operators and improve performance and efficiency. A fixture is a device that holds or supports the work piece during manufacturing operations. It enables the part to be held securely in a specific orientation, freeing the users' hands so other parts can be attached to it and necessary processes such as tightening carried out. Fixtures can also be used to hold tools supporting their weight so the operator only needs to ensure their position relative to the product and not take the weight.

A jig is a device that is pretty similar to a fixture, but also provides support in the processing operations by guiding cutting tools. The complexity and usefulness of fixtures varies, in some cases a simple device locking the part to the table top is sufficient; however, for more complex heavier products a much more sophisticated fixture is necessary, with additional capabilities, such as the ability to rotate, etc.

A number of considerations should be taken when designing fixtures to ensure they are optimizing the operator's capabilities both physically and cognitively, as they can play a significant role in providing the operator with cues and clues. The alignment of fixtures on the workstation should correspond to the order the assembly tasks should be carried out as well. The alignment of fixtures should also take into the consideration the way in which the material will be supplied to the workstation, so as to reduce the time spent orienting the material. By having a fixture that determines how the product should be orientated, the need to recall details from memory is reduced, which is particularly beneficial for operators who work on several product variants. The shape of fixtures is often a negative form of the part or component that needs to be assembled so also acts as a device to aid the assembler.

5.11. Kitting

Kitting is a method where all the required components necessary to make a product or subassembly are delivered to the operator's workstation inside a container called a kitting bin. The container often uses templates or is structured in such a way that the components can only be stored one way. Having a structured layout provides support for the assembler indicating in which order the parts should be removed and assembled, while supporting the kitter by visually showing what parts are required and in what quantity. The kit also acts as a memory trigger or early warning symbol because if the box is not empty when the worker has completed their task it is clear they have made an error somewhere during the assembly. While the value of this technique has been questioned from a materials handling viewpoint, as we will see in a later chapter, there is no doubt that from a cognitive perspective it benefits the operator.

5.12. Standardized work

Standardized work is a key part of lean manufacturing philosophy; it stops everyone from taking the "this is my way of doing things" approach and rather provides an optimized standard method that all workers should take (assemblers, machine maintenance, managers, etc.). This method means workers don't need to choose between numerous possible ways of completing the task, rather there is only one clearly defined way, the best way. By providing workers with a specific set method to carry out tasks, over time the process will become engrained in their memory, reducing the time and energy associated with memory recall. By combining all the different elements of the worker's task into a sequence, efficiency and productivity can be achieved as well as cognitive support for the worker. The use of other methods such as kitting contributes to standard work as the material is presented in a certain order, based on the standardized way the part should be assembled. Standardized work not only applies to the necessary sequence of tasks the assembler should conduct; it also applies to the state of the workstation. So pictures are often displayed showing what the normal condition of the workstation is and how it should be left and the end of a shift. In Toyota's Total Quality Management Philosophy, having standardized processes is key, as it provides the baseline needed to facilitate continuous improvement (kaizen-implementation of incremental change) (Womack, 1996). Workers are encouraged to identify potential areas of improvement that could become the new standardized procedure, which helps to create a satisfying and fulfilling work environment. Standardized work generally involves a high level of documentation. This can be particularly beneficial for training purposes, making it easier for new personnel to get to grips with quickly.

5.13. Work instructions

In its simplest form, a work instruction provides the operator with written guidelines or pictures of how the part should be assembled. Some work instructions can be quite open, only specifying the key distances or torque required with little guidance on the specific details of how the operator should actually perform the task. Other instructions utilize standardized work principles, ensuring operators are aware of the only correct way of implementing the necessary tasks. Instructions can be provided in paper form or through specialized training; however, the recent trend is for production facilities to have computers and screens located at the workstation. These provide operators with information and the necessary instructions (both text and pictures) as to how parts should be assembled. The operator has access to all the parts stored on the system and can obtain the necessary information by entering the part identification number. In some systems instead of manually typing in the part identification number to view the instructions, the operator simply scans an ID card and instructions for the part in question are provided; this method contributes to quality control as all defects can be traced.

More complex systems utilize picking by light. Initially the user is guided to the necessary material, then for assembly operations a light ball is situated where the production step is carried out and is illuminated when necessary. A sensor then picks up the assemblers presence and provides them with a current work instruction on the display screen. Only when the task has been correctly carried out can the assembler move onto the next step. For instance if only five screws have been mounted instead of the required six the system won't allow the operator to conduct the next step and an alarm will sound, alerting them of their error. This method also limits the need for operators to spend time and energy retrieving information from their memory or trying to correctly interpret a scenario. This is particularly valuable in environments where a high number of similar product variants exist, providing operators with the correct level of support. Such systems can be used anywhere and by operators of any nationality as the onscreen instructions can be in several languages. This ensures that a standardized way of work is followed throughout the whole company regardless of the geographic location of the different sites. Using a software based system also means that should any modifications to the assembly instructions need to be made; the system can be updated with no hassle with changes being made to all stations on the line simultaneously.

5.14. Poka yoke

A number of mistakes in production leading to defects and reduced quality are a result of assemblers simply forgetting to do something. Poka yoke was introduced as an attempt to combat this issue, eliminating defects by correcting or alerting humans of their errors as soon as they occur. *Poka yoke*, a term that originated in Japan, means "mistake proofing" and is concerned with preventing errors from becoming defects before the fact. Many production facilities purposely implement tools, equipment or procedures for error proofing, making it very difficult for mistakes to be made. By only providing one way of holding or storing the part, both kitting containers and fixtures at the workstation act as poka yokes.

Pick by barcodes

This method utilizes barcodes and an optical barcode scanner. A terminal provides the operator with real-time data collection information about where they need to go, what they need to pick and in what quantity, using either text or images. The operator uses the device to scan the barcode on the storage box and the terminal provides them with information regarding the desired quantity. The barcode scanner and terminal are either handheld, secured around the lower arm, or truck-mounted. This system tends to be more cost-effective than pick by light in lower volume environments. However, unlike other picking systems, the operator needs to look at the screen to retrieve the necessary information, which can be an inconvenience. This system is considered to be one step up from using a paper sheet to carry out picking tasks; however, it is not suitable for certain work environments when operators need to wear protective clothing such as gloves.

Pick by light

This method uses lights positioned on shelves, flow racks or work benches to direct and indicate to the operator what they should do next. At the right point in the sequence, the light will guide the operator to a certain location. Once they have completed the task the light will either go off automatically based on sensors or the operator will manually confirm the action by clicking the illuminated button, triggering the next light in the sequence to illuminate. In addition to a light, some systems are fitted with a display showing the necessary quantity or other information.

Despite being called pick by light, this method is not limited to picking. It can also be used to provide information about assembly tasks, for instance which tool should be used and what torque should be applied. The system can also indicate the correct storage container for items to be placed in after assembly ("put to light"). This system is considered more user-friendly than picking by barcode, as the operator's hands are kept free. Many argue that this system is the fastest picking method, as users don't need to refer to a screen or wait to hear instructions; rather, their attention is instinctively drawn towards the light. Should changes be made to the assembly line, the light modules can be easily moved and updates made to the software infrastructure.

Pick by voice

This system is similar to pick by light but uses the sense of hearing to gain the operators' attention, rather than lights. Each operator wears a headset and is provided with the necessary information to know what to pick, in what quantity and where it is located. In this method both the user's hands and eyes are free. To confirm the pick, the operator can use voice control where they will repeat some of the product information (e.g. the last four digits of the barcode), or a sensor positioned in the container will detect their selection. Unlike pick by light, this technique can be used even when multiple operators are working in the same area. The use of both pick by voice and pick by light make it relatively easy for new workers to learn their new work tasks quickly.

Andon

Andon systems provide a visual display that all workers can see to show the status of the plant floor. Enhanced visualisation is said to not only create a sense of belonging in teams, but also point out when problems in the process occur, alerting management, maintenance and other workers down the line who depend on the affected station. Empowering operators to stop the production processes encourages an immediate response, which in turn should enhance the overall quality and reduce waste (Alzatex, 2014). Generally the worker at the directly affected station pulls a cord triggering an alarm or flashing lights to alert the rest of the workforce that a problem has occurred; this can also be automated. Once the issue has been resolved, the andon is deactivated so that work can continue as normal. Many industry facilities have andon coaches whose role is to resolve any issues as soon as they arise.

Study questions

Warm-up:

Q5.1)   Name the five senses.

Q5.2)   Did you use your long-term or short-term memory to answer the question above? Explain why.

Q5.3)   Name three key design factors for designing visual information.

Q5.4)   What are the four main cognitive processes that a workplace design can support?

Q5.5)   Using the SRK model by Rasmussen, explain the difference between how a novice and an expert process information when performing a task.

Q5.6)   How do poka yokes help to support human cognitive abilities?

Look around you:

Q5.7)   Imagine (or even better, visit) an airport, train station or bus terminal. What visual, sound and tactile cues are there to help people know where to go and what to do in order to start their journey?

Q5.8)   Habits matter when it comes to sensory stimulation — some signals are so familiar that our brains may have learned a routine to not interpret them as new information. Reflect on which sounds you are able to ignore and which ones shift your attention while you are working — strangers talking around you in a café? Listening to music? Beeping noises? Someone calling your name? Birdsong? The clinking of glasses and plates? The sound of crashing glass?

Connect this knowledge to an improvement project

• List the main sensory information sources (vision, sound, touch, etc.) and consider whether the worker could use other senses to be alerted to the status of the system.
• Also consider whether any sensory inputs, or a combination of them, risk to overwhelm or confuse the worker.
• Identify the tasks that need to be performed to a certain quality level – what conditions would be optimal to reach this quality level?
• Use principles of cognitive abilities and limitations to design aids to the work, such as instructions, guides, signals and fixtures.

Connection to other topics in this book:

• Good cognitive ergonomic design of a workplace can help to improve the quality and efficiency of operations, usually by decreasing the occurrence of errors and waste of material and time – this leads to good economics (see Chapter 11). Therefore, it is a good idea to measure the status of these improvement potentials and losses before and after a cognitive ergonomics improvement, to gather evidence of how much improvements can make a difference from a cost perspective. This will also help the workplace improver to formulate a good business case.
• Environmental factors (Chapter 12) are in themselves stimuli of human senses that can confuse or overwhelm the human at work. Taking these factors into account should always be done alongside considerations of making cognitive ergonomics improvements, and care should be taken so that adaptations to the environment (e.g. using gloves, protective equipment, etc.) do not hinder the human's cognitive abilities.

Summary

- There is a need to design workplaces for workers' mental capacities, as well as their physical.
- Workplace designs that lessen the impact of fatigue can help to decrease unnecessary mental workload and avoid hazardous accidents.
- The brain, aided by the senses, processes and interprets information from the environment, enabling decisions to be made.
- Cognitive abilities are a combination of skills, experience, pattern recognition, attention, memory, ability to focus, expectations and associations.
- Vision is connected to perception, with the human mind always looking for patterns and structure that can be determined as meaningful.
- Contrast, colour intensity and strength of lighting all affect a human's ability to take in visual information.
- Sound complements vision – particularly in environments that can overload us with visual stimuli, sound can be used as a warning for workers.
- Cognition is a combination of sensory stimulation, focus, perception, working memory, long-term memory and interpretation.
- Memory enables information, experience and rules to be stored in the brain.
- Short-term memory allows us to recall recent events but is limited in how many information points it can store, typically 7 ± 2 chunks of information.
- After information has been processed, it can be stored in the brain's long-term memory, which has enormous capacity.
- Recalling information from the long-term memory is easy for frequently occurring events, but cues (significant signals) are often required to stimulate recall of events from long ago.
- The 13 design principles introduced should be considered when designing to support attention, perception, memory and mental models.
- Tools such as DFA, standardized work, fixtures, kitting, poka yoke, picking aids and andon should be used in industry to support workers.

Notes

[1] Also known as the "reptile brain".
[2] Knowing the position of parts of our body in space and what condition our muscles are in, i.e. whether they are contracted or not. This is also called *proprioception*.

5.15. References

Alzatex – Lean Timers. (2014). What is Andon? [Online]. Available from: http://lean-timer.com/lean-manufacturing-andon/ [Accessed 15 Jan 2014].

Bohgard, M. (Ed.) (2009). *Work and Technology on Human Terms*. Stockholm: Prevent. ISBN 978-91-7365-058-8

Boothroyd, G., Dewhurst, P. & Knight, W. (2002). *Product Design for Manufacture and Assembly,* 2nd edition. New York: Dekker.

Kroemer, K. H. E. & Grandjean, E. (1997). *Fitting the Task to the Human: A Textbook of Occupational Ergonomics.* London; Bristol, PA: Taylor & Francis.

Hart, S., & Staveland, L. (1988). Development of NASA-TLX (Task Load Index): Results of empirical and theoretical research. In P. Hancock & N. Meshkati (Eds.), *Human Mental Workload* (pp. 139–183). Amsterdam: North Holland.

NASA. (n.d.). NASA TLX: Task Load Index [Online]. Available from: http://humansystems.arc.nasa.gov/groups/tlx/ [Accessed 8 Jan 2014].

NASA. (2014). NASA Task Load Index [Online]. Available from: http://humansystems.arc.nasa.gov/groups/TLX/downloads/TLXScale.pdf [Accessed 9 Jan 2014].

Rasmussen, J. (1983). Skill, Rules, Knowledge: Signals, Signs and Symbols and Other Distinctions in Human Performance Models. IEEE *Transactions on Systems, Man and Cybernetics* (SMC-13) 3:257–266.

Reason, J. (1990). *Human Error.* Cambridge: Cambridge University Press. ISBN-0-521-31419-4

United States Nuclear Regulatory Commission (2013). Backgrounder on the Three Mile Island Accident. [Online]. Available from: http://www.nrc.gov/reading-rm/doc-collections/fact-sheets/3mile-isle.html [Accessed 8t Jan 2014].

Womack, J. P. & Jone, D. T. (1996). *Lean Thinking: Banish Waste and Create Wealth in Your Corporation.* Michigan: Simon & Schuster.

Wikimedia Commons. (2014a). File:Kanizsa triangle.svg [Online]. Available from: http://commons.wikimedia.org/wiki/File:Kanizsa_triangle.svg [Accessed 9 Jan 2014].

Wikimedia Commons. (2014b). File:Mond-vergleich.svg [Online]. Available from: http://en.wikipedia.org/wiki/File:Mond-vergleich.svg [Accessed 9 Jan 2014].

CHAPTER 6

Psychosocial Factors and Worker Involvement

THIS CHAPTER PROVIDES:

- Specific psychosocial factors that influence the human's ability to perform and develop.
- Positive and negative effects of stress, task demands and control over the work.
- Different arguments for using models of a workplace to involve users and other stakeholders in participative ergonomics design.
- Characteristics of a psychosocially healthy workplace.

How to cite this book chapter:
Berlin, C and Adams C 2017 *Production Ergonomics: Designing Work Systems to Support Optimal Human Performance.* Pp. 107–124. London: Ubiquity Press. DOI: https://doi.org/10.5334/bbe.f. License: CC-BY 4.0

WHY DO I NEED TO KNOW THIS AS AN ENGINEER?

Human beings in a workforce are not just a physical and cognitive work resource; they are also individual personalities whose performance is affected by their psychological well-being, motivation and subjective experience as an employee. They are also team members trying to navigate social codes and expectations, and furthermore they are a private person outside of the workplace.

In order for workplace design to create the best possible conditions for human workers to perform well, it is important to understand some human psychological reactions, stress tolerance levels, motivating mechanisms, support needs and the need to have influence on how they work. We collect all these aspects under the umbrella term "psychosocial factors". While there are still workplaces today that assume that the human workforce is there to obey instructions blindly, a socially sustainable workplace cannot ignore the importance of interplay between humans, and allowing them opportunities to engage themselves and affect the work that they do. The branch of ergonomics known as macroergonomics concerns itself with the influence of organizations on ergonomics and the interplay between humans, rather than the very specific design of equipment and technology interfaces that characterize human-machine interaction, or microergonomics.

One important aspect of macroergonomics is participative techniques, where different methods are used to involve system users, workers and other stakeholders to engage with their knowledge in making changes to the workplace or giving opinions and ideas in the system design process. Such a workplace has a better chance of attracting and retaining staff for a longer time, at the same time allowing them to become valuable, experienced knowledge resources in the production system.

While it may seem unusual for engineers and workplace designers to care about psychosocial work environment, it is beneficial from a systemic point of view to know how teamwork and human motivation is impacted by the work environment, and how the contents of the previous chapters (physical loading, cognitive ergonomics) are interrelated with effects on the human psyche. In the end, all of these aspects interact to impact the human's ability to perform in the workplace. Focus on this aspect is increasingly understood as part of creating responsibly run workplaces. In 2015, the Swedish Work Environment Authority recently issued a legal provision placing responsibility for organisational and psychosocial work environment on the employer, which means that it is crucial for management roles to grasp what is within their scope of control to ensure psychosocial health.

WHICH ROLES BENEFIT FROM THIS KNOWLEDGE?

The *manager/leader* is the primary benefactor of this chapter's knowledge. Knowing which psychosocial stressors and risks are present in work and the workplace is an important precursor to making sustainable long-term decisions about staffing, task allocation, training and competence development, and worker well-being initiatives. The knowledge can be beneficial to building and supporting the growth and performance of teams.

The *system performance improver* can benefit from an understanding of how the tangible aspects of workplace design are connected to needs and limitations of humans in a social context. The book has so far covered performance aspects on an individual level, but this chapter introduces aspects of teamwork, hierarchy and decision latitude, all of which can be directly supported or hindered by physical and cognitive loading in the workplace.

The *work environment/safety specialist* benefits from knowing the overall management perspective and the challenges that they face from a personnel-management point of view. Worker safety risks in a psychosocial sense may be difficult to recognize and target without sufficient knowledge of the delayed reactions humans may exhibit to chronic stressors and demotivation; therefore, it is crucial that this work role is able to recognize these risks in their latent state and alert management to the possible consequences.

6.1. Macroergonomics

As mentioned before in Chapter 1, the scope of ergonomics has undergone several generations of "widening" its areas of application. The developments in the 1980s directed the field's attentions to the social and organizational context of ergonomics – in a word, macroergonomics. No two organizations are alike, which means that the awareness, support, understanding and emphasis on creating better workplaces varies a lot across company sizes, industrial sectors, history, geographical location, cultures (particularly regarding hierarchy and influence) and the current ideals of the times. Understanding that these social contextual factors can facilitate or hinder positive improvement developments is a central tenet to understanding the meaning and impact of macroergonomics.

Techniques for understanding macroergonomic factors include interview studies, organizational questionnaires, field studies, focus groups, etc. – in other words, there is a lot of emphasis on studying the views and agendas of different human actors, both as individual actors and as teams. Hendrick and Kleiner (2001) describe macroergonomics as being not only top-down (strategic, where leadership states that improvement is a mission), but also "bottom-up" (participatory, where workers get involved) and "middle-out" (focusing on processes).

In contrast, the type of design challenges this book has covered in previous chapters is sometimes labelled microergonomics; this is when the scope of concern is focused on improving the human-machine and human-process interface on the basis of human needs and capabilities. However, there is really no other reason to use that label unless it is necessary to make a contextual distinction between that and macroergonomics.

6.2. Psychosocial environment

The term psychosocial gained recognition in the late '70s as an important aspect of healthy work environments. It was recognized that human beings are active in, and react to, their immediate surroundings. Insomuch as work processes and work environments affect them, humans also have some level of control and influence over them. This happens on both a psychological level, affecting our thoughts and feelings, but also in a biophysical sense (particularly in the long run) where the psychosocial environment can affect our hormone levels, posture, ability to concentrate, metabolic processes, sleep patterns, etc.

There are many dimensions to the psychosocial environment. First of all, there is the social aspect of working in teams, which may be more or less functional depending on whether the team can accept each other, communicate and collaborate. Surrounding this is the cultural aspect, which dictates the pace of life in that part of the world (in a geographical and time-related sense), and to some degree the order of priorities in life; cultural influence varies depending on where in the world we are, what type of sector we work in, and the changing times. Finally, personal lifestyle as part of the psychosocial environment dictates the individual's balance between work and leisure time, and what is considered a satisfactory quality of life (regarding income, personal involvement at work, opportunities for development and empowerment). This tends to change with the times we live in, but also with different stages of life that the individual goes through.

Certain psychosocial factors can be analysed separately in order to deliberately design the best possible psychosocial conditions in the workplace, as far as this is possible for the employer to control. The aim is to create a workplace that is stimulating, motivating, supportive and sufficiently rewarding for the workforce, hopefully resulting in engagement, creativity, company loyalty and increased competence as a result of employees wanting to stay longer.

6.3. Positive and negative stress

Any time that the mind and body are engaged to perform a task to meet time, quality or performance demands, our alertness increases and we are biologically prepared to react (see Chapter 5.4).

Whenever we feel that a situation is stressful, exciting, alarming or the like, regardless of the perceived consequences, it is possible that the body psychosomatically interprets this as danger, releases stress hormones and automatically enters "fight or flight" mode – a condition stemming from humanity's caveman days, where sudden threats of danger required either a fighting response or a quick escape from the danger. What the body does under stress is to release hormones from the adrenal gland, particularly adrenalin and noradrenalin (Kroemer and Grandjean, 1997) and redistribute how the nutrients in the body are used, to prioritize muscular response. The heart rate and breathing rate increase and the senses become more acute, but the body directs resources and nutrients away from processes like digestion, regrowth, learning and the immune system.

When we are in a challenging situation but confident of being able to complete the task successfully, the adrenaline-kick is only temporary and can be called positive stress, since it serves to increase our alertness and stimulate us with a manageable challenge[1]. Once the task is completed we experience a drop in adrenaline and probably a sense of success that may be chemically reinforced. However, in situations where we feel that we are unable to succeed, and especially if they occur very frequently, the stress becomes negative stress. If we spend large amounts of time in states of negative stress, never letting the adrenal hormone levels fall again, we run the risk of suffering chronic stress symptoms such as ill-health, an overworked heart, anxiety, muscle tension, digestive problems, high blood pressure, exhaustion and weakened capacity to repair and recover. For this reason, it is essential to remember that workplace stressors are not just caused by obvious time restrictions; we may also be stressed by high demands, bad communication, emotional triggers and relational malfunctions like conflicts and interpersonal irritations.

6.4. Boredom

Another important aspect of workplace psychosocial health is boredom – the mental state that occurs when the level of stimuli in an environment is perceived as low and monotonous enough for an individual to stop concentrating at the task at hand, usually as a result of a mismatch between the task demands and the competence or skill level of the individual. The negative consequences of boredom include deactivation of higher nervous centres in the brain, feelings of weariness, and lack of alertness that may lead to quality deficiencies or errors.

The most negative state of boredom (from a motivational and alertness point of view) occurs when the task is not monotonous enough for the worker to think about other things entirely if attention is slipping in and out of concentrating on the task because it is not entirely internalized as a routine skill, the worker may feel frustrated. Vigilance, or sustained attention, is a taxing mental state for most humans, especially if stress is part of the work situation. A related, purely emotional tension may occur when the worker feels inner conflict about whether they wish to continue performing the task to the set requirements, or whether they want to be done with it. This emotional tension may over time lead to job dissatisfaction and a deliberate decrease in performance quality.

To counteract boredom, the following points are worth considering in task and workplace design:

- Carefully match the level of the worker's competence with the difficulty of the task.
- Encourage alertness and opportunities for recovery, to make sure workers feel fresh and ready to work – fatigue in itself can exacerbate boredom.
- Avoid work conditions that can increase boredom: solitary work with no contact between colleagues; dim lighting; too-warm climate; very brief and repetitive work cycles; too many non-critical alerts that do not require decisions or action.
- Learners are often more content to do a simple task while they are still in a learning process.
- Design a learning scheme into the tasks, perhaps by "unlocking" increasing levels of difficulty.

(Adapted from Kroemer and Grandjean, 1997 p. 220)

To add some nuance, some recent scientific results from the field of psychology (Gasper and Middlewood, 2014) have re-evaluated boredom, seeing it as a source of creativity and a needed window for daydreaming and reflection in a world that is increasingly distractive and stressful. The study showed that test participants who were bored outperformed stressed participants in creative thinking. However, for our purposes of designing a production workplace, we are targeting the kind of distractive boredom that can lead to slips, errors and mistakes.

6.5. Motivation

Generally, *motivation* can be defined as the mental state where a task or overall goal carries meaning for the person performing it, which increases their willingness to take action to complete specific goals.

It is useful to distinguish between the reasons that motivate an individual to act or pursue a goal, or even accept certain conditions. When a task is perceived as meaningful in itself and the individual voluntarily applies effort and time, this is called intrinsic motivation. When the task in itself may not be enough to motivate a person to do something, there may be other reasons – such as a reward, a higher overall goal (of which the task is a step on the way), a sense of developing skill and self-actualization, or getting recognition for the effort or achievement. Such external motivators are called extrinsic, and may cause a person to put up with some discomfort or inconvenience to complete the task, because the end result of completing it brings the person closer to an overall goal. Some of these goals may relate to human *needs,* which are compelling physiological and psychological drives to survive, thrive and self-actualize.

A classic and well-known model for the hierarchy of human motivational factors is Maslow's (1943) Hierarchy of Human Needs (Figure 6.1), which is most often illustrated as a pyramid, and explained (in Maslow's words) as "When the most prepotent goal is realized, the next higher need emerges" (p.370).

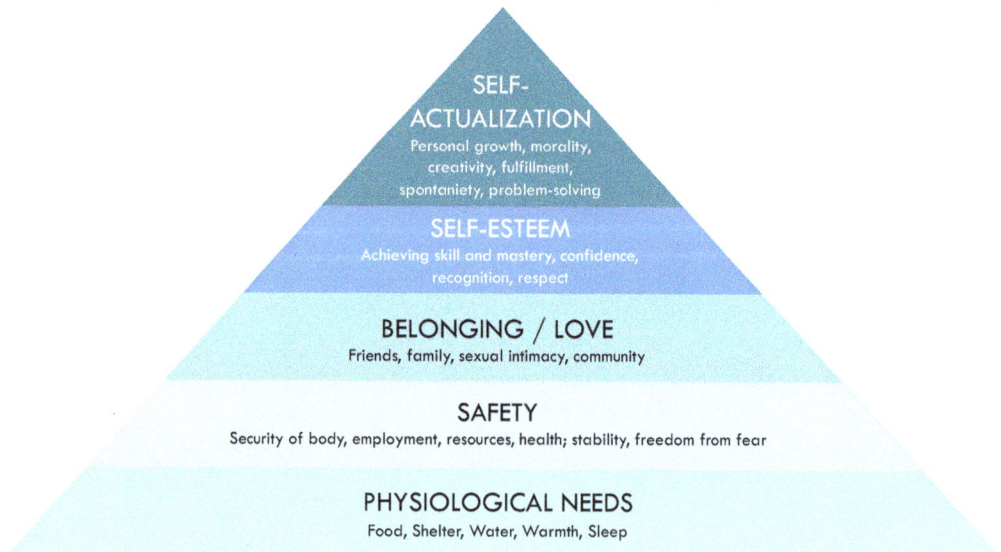

SELF-ACTUALIZATION
Personal growth, morality, creativity, fulfillment, spontaniety, problem-solving

SELF-ESTEEM
Achieving skill and mastery, confidence, recognition, respect

BELONGING / LOVE
Friends, family, sexual intimacy, community

SAFETY
Security of body, employment, resources, health; stability, freedom from fear

PHYSIOLOGICAL NEEDS
Food, Shelter, Water, Warmth, Sleep

Figure 6.1: Maslow's hierarchy of human needs (adapted from Maslow, 1943).

Illustration by C. Berlin, based on Maslow (1943).

Although this model has been debated and updated in various instalments, its historical and cultural impact can be considered immense on the general public's mental model of human needs and drives. It is shown here mostly for cultural reference.

An alternative classification of human needs has been presented by the Chilean researcher and activist Manfred Max-Neef (1992), who defined nine basic human needs and stated that a) they are *not* hierarchical, b) *not* substitutable, and c) that they do *not* vary between cultures. Each human need is equally important for a human being to be healthy. Table 6.1 lists the nine basic needs, which in turn can be fulfilled by *satisfiers* – these are ways of *being, having, doing* or *interacting* that contribute to addressing human needs. According to Max-Neef, a non-fulfilled need of any kind is a form of human poverty.

In a work design context, it is safe to say that certain motivational factors that appeal to the needs concerning our means for survival – in modern industrialized terms, our livelihood. Certain factors must be in place for anyone to even consider taking on a task or job and staying engaged. These basic conditions are known as *hygiene factors* and include basic remuneration (payment) and guarantees for well-being such as appropriate salary level, work hours, recreation opportunities, development opportunities, social contact, etc. Some of these hygiene factors may vary across cultures, ages, stages in life and levels of skill or education. Sadly, some living conditions in the world are so desperate that in order to make a living, workers will accept high levels of danger to their safety and health in order to make a livelihood – sometimes at the cost of debilitating injuries that may limit their future ability to work and earn a livelihood, or a loss of human rights (such as having travel documents confiscated).

6.6. Psychosocial factors coupled to tasks

Although many measurement methods exist to somehow quantify stress levels, motivation, engagement, etc., many of them become an uncertain basis for changes because the reasons for experiencing stress vary from individual to individual and across ages depending on their personal life situation, education level, stage in a learning process, experience with the current tasks at hand, relations with and acceptance from colleagues, etc. It may be a good initiative to monitor the stress levels of a workforce for the purpose of introducing better support through design or planning of human resources, but it is important to remember that these measurements never stay static.

Attempts have been made to capture a holistic measurement of task workload, including both the task-related and the psychosocial aspects. As a refresher, Figure 6.2 reprises the NASA- TLX scale (Hart and Staveland, 1988) designed to measure workload. Many of the things asked for, although they are asked in a way that makes each answer individual and subjective (i.e. the scale is not absolute across humanity), can give work and workplace designers a good idea if the system's overall performance is in danger due to job dissatisfaction.

6.7. Demand-control-support model

Karasek (1979) studied the interaction between stress-inducing psychosocial factors and came up with a now classic model explaining how work demands and the level of worker's control over their tasks (decision latitude) influence stress levels at work. The axes of these dimensions simply designate the status of being "high or low", and the resulting four zones explain what stress-level

Table 6.1: Max-Neef's (1992) nine categories of human needs, and the four categories of satisfiers that can fulfil these needs (taken from Hitchcock and Willard, 2013 p.2).

Need	Being (qualities)	Having (things)	Doing (actions)	Interacting (settings)
Subsistence	physical and mental health	food, shelter, work	feed, clothe, rest, work	living environment, social setting
Protection	care, adaptability, autonomy	social security, health systems, work	co-operate, plan, take care of, help	Social environment, dwelling
Affection	respect, sense of humour, generosity, sensuality	friendships, family, relationships with nature	share, take care of, make love, express emotions	privacy, intimate spaces of togetherness
Understanding	critical capacity, curiosity, intuition	literature, teachers, policies, educational	analyze, study, meditate, investigate	schools, families, universities, communities
Participation	receptiveness, dedication, sense of humour	responsibilities, duties, work, rights	cooperate, dissent, express opinions	associations, parties, churches, neighbourhoods
Leisure	imagination, tranquility, sense of humour, spontaneity	games, parties, peace of mind	day-dream, remember, relax, have fun	intimate spaces, places to be alone, landscapes
Creation	imagination, boldness, inventiveness, curiosity	abilities, skills, work, techniques	invent, build, design, work, compose, interpret	spaces for expression, workshops, audiences
Identity	sense of belonging, self-esteem, consistency	language, religions, work, customs, values, norms	get to know oneself, grow, commit oneself	places one belongs to, everyday settings
Freedom	autonomy, passion, self-esteem, open-mindedness	equal rights	dissent, choose, run risks, develop awareness	anywhere

Note that these satisfiers are general, not targeted at production environments, but they are important from a holistic personnel-health point of view.

effects their combination may have on workers. Figure 6.3 shows the four zones of psychosocial health, describing them as "active, low-strain, passive, high-strain" in the order of increasing psychosocial risk.

A later version of this model was developed by Karasek and Theorell (1990) where an additional dimension was mapped: that of *social support*, a factor that can help stressed workers manage the job strain. Figure 6.4 shows this more nuanced model as a three-dimensional representation.

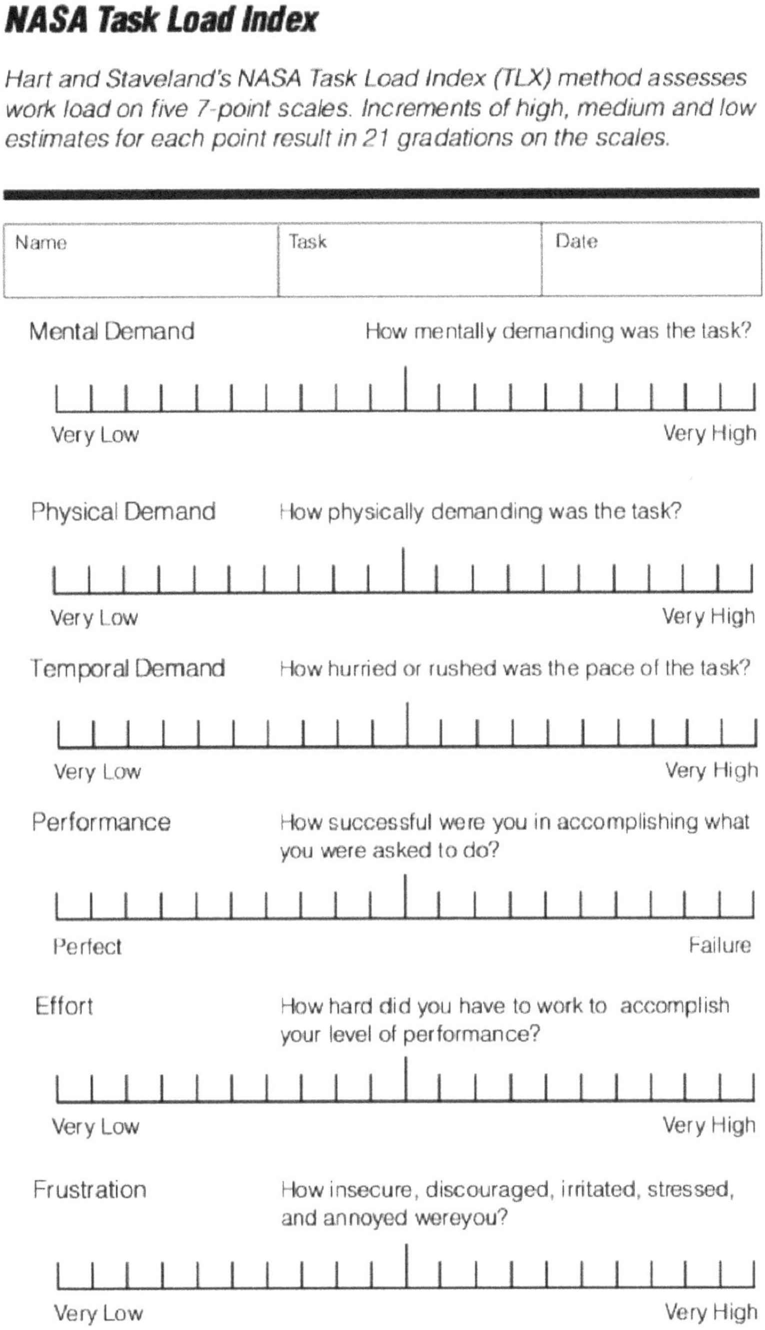

NASA Task Load Index

Hart and Staveland's NASA Task Load Index (TLX) method assesses work load on five 7-point scales. Increments of high, medium and low estimates for each point result in 21 gradations on the scales.

Name	Task	Date

Mental Demand   How mentally demanding was the task?

Very Low   Very High

Physical Demand   How physically demanding was the task?

Very Low   Very High

Temporal Demand   How hurried or rushed was the pace of the task?

Very Low   Very High

Performance   How successful were you in accomplishing what you were asked to do?

Perfect   Failure

Effort   How hard did you have to work to accomplish your level of performance?

Very Low   Very High

Frustration   How insecure, discouraged, irritated, stressed, and annoyed wereyou?

Very Low   Very High

Figure 6.2: The NASA-TLX form with its six sub-scales (NASA, 2014).

Source: NASA Ames Research Center; used with permission.

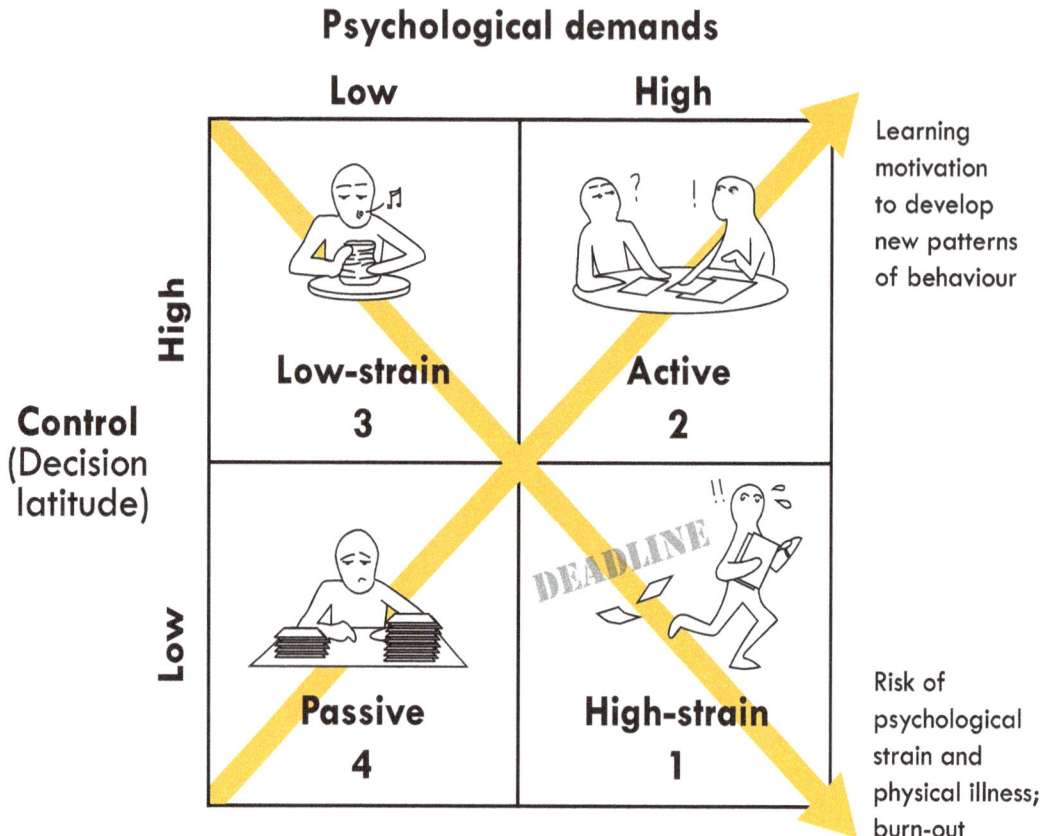

Figure 6.3: The relation between demand and control (decision latitude), adapted from Karasek (1979). Illustration by C. Berlin, based on Karasek (1979).

6.8. Participatory ergonomics

One of the most central improvement techniques from the macroergonomic approach is participatory ergonomics (also known as *participatory design*[2]) defined by Wilson as "the involvement of people in planning and controlling a significant amount of their own work activities, with sufficient knowledge and power to influence both processes and outcomes in order to achieve desirable goals" (1995 p. 37).

6.9. A process for participatory design

Vink et al. (2005) described the participatory design process as consisting of six steps, as shown in Table 6.2.

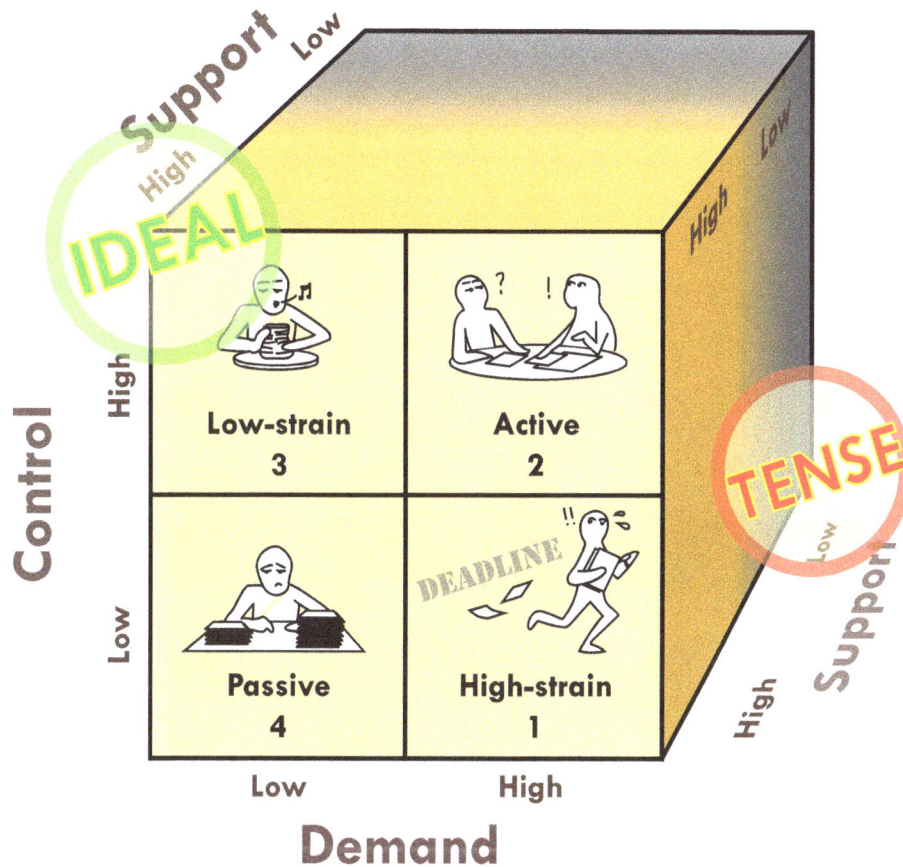

Figure 6.4: The relation between demand, control (decision latitude) and support, as theorized by Karasek and Theorell, 1990.

Illustration by C. Berlin, based on Karasek and Theorell (1990).

6.10. Using models of the design solution

One very effective strategy for eliciting discussion and feedback from the participants is to use different types of models representing the work system as a basis for discussion. Having a visible representation of the workplace layout and how the new design solution fits into it helps to direct the attention of the participants towards discussing design rather than general well-being aspects. This can be a great tool for designers to leverage not only good ideas, knowledge and suggestions, but also to encourage acceptance for the final solution among the end-users. Having a visual representation of the workplace offers an opportunity to point to specific details and relate some of the feedback to human movement and the dimensions of the human body.

Table 6.2: The participatory design process, adapted from Vink et al. (2005) and Kuijt- Evers (2006).

1) Preparation	The stakeholders are informed of the planned change project and its overall goals. The stakeholders may include end users, management, designers, specialists, operators, maintenance personnel, etc. The overall strategy for how to involve them and turn their feedback into a solution is discussed.
2) Analysis of tasks, work and health	A baseline for the design is established by studying the current practices, needs, problems and solutions in the context of the workplace. This can be achieved using observations, interviews (group or individual), simulation or questionnaires. The purpose here is not to influence, but to study how things are done.
3) Selection of improvements and design	A requirement specification for solving the identified problems and meeting the identified needs is created. This should build on user requirements and wishes. This is a good stage to involve the users in a participatory process, allowing them to engage in a forum where they can suggest ideas and improvements. When this input has been collected, new design ideas can be tested and made. A good way to make more involvement possible is to build models of the new design proposals, either in 2D or 3D format, for the participants to relate to in discussions.
4) Pilot study with the improvements	This is the stage at which testing occurs on the basis of the design models – they can be tested in the context of the real, existing workplace, in a "clinical" setting to direct attention away from details that shouldn't be the subject of feedback, or a mix between these environments.
5) Implementation	After one or more iterations of steps 3–4, the new design can be implemented in its real context. The participants are informed and educated about the implications of introducing the new design.
6) Evaluation	After an adjustment period where the end users get accustomed to the new solution, an evaluation can be carried out to determine if well-being and system performance have increased compared to the baseline established in step 2. If found necessary, this participatory evaluation can become the basis for further improvements.

One important thing to remember is that the detail level of the model will determine the level of feedback gained from the participants; for example, if you do not want feedback on small details like the exact size and shape of buttons, but rather the heights, depths and layout of work areas and tables, then it is possible to temporarily omit those details in the model, explaining to the participants what type of feedback you are expecting. Another useful technique is to ask the participants to imagine a scenario where they are trying to complete a task in the new design solution. This type of imaginary goal can help the participants direct their feedback towards things the designer cannot know or guess, such as experiences, work procedures, anecdotes, safety concerns and workarounds. Different model representations have different pros and cons, as follows:

2D drawings

Two-dimensional drawings constitute a rather common representation of new workplace designs. Quite frequently they are shown as technical drawings from the top or side view, describing the layout. Although there may be a cultural expectation that these drawings are easy to understand, there is a risk that the bare-bones flat representation on paper or screen does not allow the users to evaluate all aspects of working in the new design solution. Some pros of 2D drawings include the ease of distributing the information to all different participants, including the ability to mail or send them to faraway participants, and the fact that writing and drawing on these representations allow individuals to comment and suggest changes rather easily. However, it is difficult to get a fair representation of heights, depths, distances and the relation of these dimensions to the human body. Also, a 2D drawing may seem like a finished architectural blueprint that does not encourage workers to suggest further changes.

3D scale models

It is possible to build a small-scale three-dimensional representation of the design idea, using cheap materials such as cardboard, clay, foam board and glue, etc. To do this requires materials, work time and some model-building competence, and perhaps also a modelled human representation to go with it, in order for the participants to judge how sizes, depths and distances relate to the human worker's size. Three-dimensional scale models are comparatively cheap, easy to change, can be easily transported and stored, and provide very a good discussion basis for feedback in groups. Also, depending on the chosen detail level and the level of "finish" of the model, participants may feel that they have the possibility to suggest changes by building or modifying the actual model using the same materials as the designer. It is important not to intimidate participants from changing, moving or manipulating the model. At the same time, the disadvantages are that the time and effort necessary to make the model(s) might result in only one or a few being built, and they essentially demand the physical presence of the participants in order for evaluation to take place.

3D full-scale models (mock-ups)

A full-scale model of the new design has the advantage that no human representation is needed; the users themselves can relate their own bodies to the new design proposal, which is particularly useful when judging movement patterns, reach distances, lines of sight and general comfort. The visual and tactile representation of the workplace may further enrich the feedback given from participants, and elicits good feedback in group discussions. Like with 3D scale models, model building competence, materials and time are needed, but it is also possible to suggest that the model is not a finished design, by making the representation seem "rough at the edges" and open to modification.

It is also particularly important in a full-scale model to be deliberate about the level of detail shown to the participants, in order to direct attention and feedback to the design aspects that the designer wants commented. Full scale models also demand the physical presence of the participants in order for evaluations to take place – not to mention considerable space, and the license to occupy that space for some time until the evaluation is over.

Consider the following differences when choosing a type of model for participatory work:

	2D drawings	Small 3D scale models	3D life-size mock-ups	3D CAD	3D imaging
Typical focus	Overall layout, formal architectural boundaries	Overall layout, line of sight, worker movement patterns	Simulating movement of self in the environment, interaction	Intended detail design, space, tolerances	Current state of worksite, available space, collisions
Portability/ Accessibility	High – can be brought to meetings packed flat or in a roll, or in digital format	High – can be brought to users, but may be fragile	Low – need to be built on-site	High – can be shared digitally, but requires that users can navigate the model	High – can be shared digitally, but may be limited by large file size and confidentiality
Advantages	Familiar format, can be drawn and written on to give input	Portable, gives an overview, lets users point and relate to heights and lines of sight, can be easy to modify	Allows active interaction with the model; the user can relate their own body to the model's dimensions	Simple, can be a "perfect intended" model; easy to change and modify	Most exact representation of current state/space
Drawbacks	Can be abstract, formal and not so engaging; because of the top-view, it can be hard to judge heights, sizes and lines of sight	May be easily damaged depending on the material; may require a human representation to give a size reference	Requires time, space and material; users must come to the model; can perhaps be transported, but with difficulty	May deviate from real machines and interfaces; most 3D CAD models are an ideal representation	Captures all data regardless of usefulness; requires post-processing and interpretation to become meaningful data
Costs of updating	Man-hours for updating formal drawings; requires expertise and can therefore be expensive	Man-hours for re-building/ modifying; rapid; probably not expensive during prototyping phases, but requires material	Man-hours for re-building/ modifying; relatively rapid but may be physically intensive and time-consuming, and may require physical materials like plywood or cardboard	Man-hours for modifying CAD model; requires expertise; individual detail change may be rapid, but updating CAD library requires time and effort	Man-hours for inserting CAD models (rapid), *or* re-taking the scan and post-processing the point-cloud data (time-consuming); requires access to the site, which may be inconvenient and require effort and time

3D virtual models (CAD)

The current maturity of computer aided design (CAD) software allows for quick an accurate modelling of three-dimensional workplace layouts in a computer setting. This means that the 3D representation can be viewed and studied on a computer screen, and (like 2D drawings) can be very easily distributed between participants, even if they are far apart geographically. Another advantage is that it is possible to include human representations (called manikins) inside the 3D model, and they sometimes include built-in analysis tools such as being able to see out of the eyes of the computer manikin, in order to judge the line of sight. Although it is a matter of education and familiarity with computer environments, it is important to note that some users may feel intimidated by navigating an unfamiliar 3D virtual environment, so it is essential that if virtual 3D models are used, all participants must be familiar and comfortable with navigating the model. If not, the designer runs the risk of getting little or no feedback because the participants may feel reluctant to admit that they could not get a good grasp of what the model tried to convey. In guided group discussions, this can be helped by offering assistance to unfamiliar users, and encouraging than to try out different functionalities offered by the software.

3D imaging (digital)

Recent developments in 3D imaging technology has brought about a large number of new measuring equipment, e.g. structured light sensors, photogrammetry, or 3D laser scanning, that allow us to capture a 3D representation of an existing object or environment (for example, a product or an entire factory). The equipment is either active (meaning that it emits signals and registers the returns) or passive (i.e. simply captures the existing signals). The 3D imaging devices are able to capture, often with very high precision, the spatial position of surfaces found in the environment, and these surfaces are registered as "point clouds", or clusters of positioned points in a digital 3D environment with an orthogonal axis. The equipment often includes an RGB sensor that sweeps the same area and assigns every single data point with colour data, allowing us to see the exact colour and dimension of every object recorded during the 3D scanning. These digital model representations are very useful for getting consensus in a group for the size and shape of a space is that may be the target for a design change, and it is possible to place CAD objects (for example a 3D CAD model of a machine) to see whether it fits in the existing architecture. These models may be a category apart, as they serve more as a visual discussion aid for stakeholder input, but the participatory aspect of being able to change the model interactively is currently limited.

Study questions

Warm-up:

Q6.1)   What is the difference between intrinsic and extrinsic motivation?

Q6.2)   What is the benefit of positive stress, and the drawback of negative stress?

Q6.3)   Explain why chronic stress is a workplace risk.

Q6.4)   Why is it OK to carry out "boring" routine tasks when one is a beginner?

Q6.5)   What are the benefits of participatory design?

Look around you:

Q6.6)   Consider any profession within an organization that you are familiar with – what organizational mechanisms and routines are in place to provide specific workers with an appropriate level of control, demand and support in their tasks?

Connect this knowledge to an improvement project

- See question 6.6: if you are observing work in an organization, try to list the organizational mechanisms and routines in place to provide workers with an appropriate level of control, demand and support. If you see that any of these components are under- or over-dimensioned in such a way that work or teamwork is negatively impacted, list it as an improvement potential.
- Use the ideas of *demand, control* and *support* to guide face-to-face interviews with workers. Also consider asking about sources of stress, boredom, motivation and demotivation. The NASA-TLX may offer inspiration.
- Use models (in 2D or 3D) representing the workplace to discuss improvement potentials with workplace stakeholders. Use the model representation to steer the discussion of how the workplace design supports or hinders tasks. Also, use a human representation to let discussion participants show movement pathways, positioning and space requirements.

Connection to other topics in this book:

- The task analysis described in Chapter 7 is a good first step towards addressing psychosocial factors in a structured manner. Hierarchical breakdowns can make it easier to map identified psychosocial risks associated with a particular operation (such as a particularly stressful one).
- Environmental factors (Chapter 12) may be stressors in and of themselves, sometimes without workers realizing it. Sometimes a demanding environment may be a critical factor in exhaustion and burnout, particularly if combined with problematic job demands or dysfunctional teamwork and leadership.
- Aspects of socially sustainable workplaces (Chapter 13) are tightly coupled to psychosocial factors and participation. The likelihood that valuable employees want to remain with the company in the long run is usually tightly coupled to the job's psychosocial factors, particularly participation and motivation.

Summary

- A way to look at the concept of leadership in the context of designing workplaces is to define it as "creating the right conditions for other people to perform".
- A psychosocially healthy workplace strives towards low negative stress, high motivation, and the right levels of control, demands, decision latitude, support and supervision.
- Stress can be positive or negative – when positive it challenges, motivates and increases alertness; when negative it can cause physical health problems, strain, anxiety and chemical imbalance.
- The demand-control-support models by Karasek and Theorell illustrate how these psychosocial factors influence performance, engagement and well-being. They also offer an explanation for the occurrence of positive and negative stress.
- Determining appropriate levels of support and supervision are the responsibility of the leadership, and the type of leadership that is appropriate varies depending on the competence level and maturity of the individual employee.
- The right levels of control, demands, decision latitude, support and supervision encourage employees to grow both in confidence and competence.
- Some "conditions for people to perform" are very much influenced by good or bad workplace and/or equipment design, but can also be influenced by the workers' ability to accept, reject or influence those designs.
- A good technique for fostering more worker engagement and sharing of knowledge is to use participatory techniques using models of the design proposals to stimulate discussions and questioning.

Notes

[1] Sometimes, people who actively seek out stressful or exciting situations (known as "adrenaline junkies") specifically to experience the adrenaline release experience a self-induced high. However, it would be too much of a simplification to say that this is purely because of adrenaline, since other substances like endorphins (positive neurotransmitters) may also be released.

[2] In some circles, *participative design* maybe a more accepted term if the purpose is to collect more input than just ergonomics aspects, so choose what to call it based on the interests of your target audience.

6.11. References

Hart, S., & Staveland, L. (1988). Development of NASA-TLX (Task Load Index): Results of empirical and theoretical research. In P. Hancock & N. Meshkati (Eds.), *Human Mental Workload* (pp. 139–183). Amsterdam: North Holland.

Hitchcock, D., Willard, M. (2013). Confused about social sustainability? What it means for organizations in developed countries. [Online] Available from: http://www.sustainabilityprofessionals.org/sites/default/files/Confused%20about%20social%20sustainability_0.pdf [Accessed 24 Feb 2015].

Gasper, K., Middlewood, B.L. (2014). Approaching novel thoughts: Understanding why elation and boredom promote associative thought more than distress and relaxation. *Journal of Experimental Social Psychology, 52*, 50–57. ISSN 0022-1031. http://dx.doi.org/10.1016/j.jesp.2013.12.007.

Karasek, R. (1979) Job Demands, Job Decision Latitude, and Mental Strain: Implications for Job Redesign. *Administrative Science Quarterly, 24* (2), 285–308.

Karasek, R. & Theorell, T. (1990). *Healthy Work: Stress, Productivity, and the Reconstruction of Working Life*. New York: Basic Books.

Kuijt-Evers, L. (2006). Comfort in using hand tools: Theory, design and evaluation. S.l: TNO Kwaliteit van Leven, Arbeid.

Maslow, A. H. (1943) A theory of human motivation. *Psychological Review, 50* (4): 370–396.

Max-Neef, M. (1992). Development and human needs. In: Ekins, P., Max-Neef, M. (Eds.), *Real Life Economics*. London: Routledge. 197–214.

NASA. (2014). NASA Task Load Index [Online]. Available from: http://humansystems.arc.nasa.gov/groups/TLX/downloads/TLXScale.pdf. [Accessed 9 Jan 2014].

Vink, P., Nichols, S. & Davies, R. C. (2005). Participatory ergonomics and comfort. In: *Comfort in Design: Principles and Good Practice*. Vink, P. (Ed.). Boca Raton: CRC Press. 41–54.

Wilson, J. R. (1995). Solution ownership in participative work redesign: The case of a crane control room. *International Journal of Industrial Ergonomics, 15* (5): 329–344.

PART 2

Engineering the System
around Humans

CHAPTER 7

Data Collection and Task Analysis

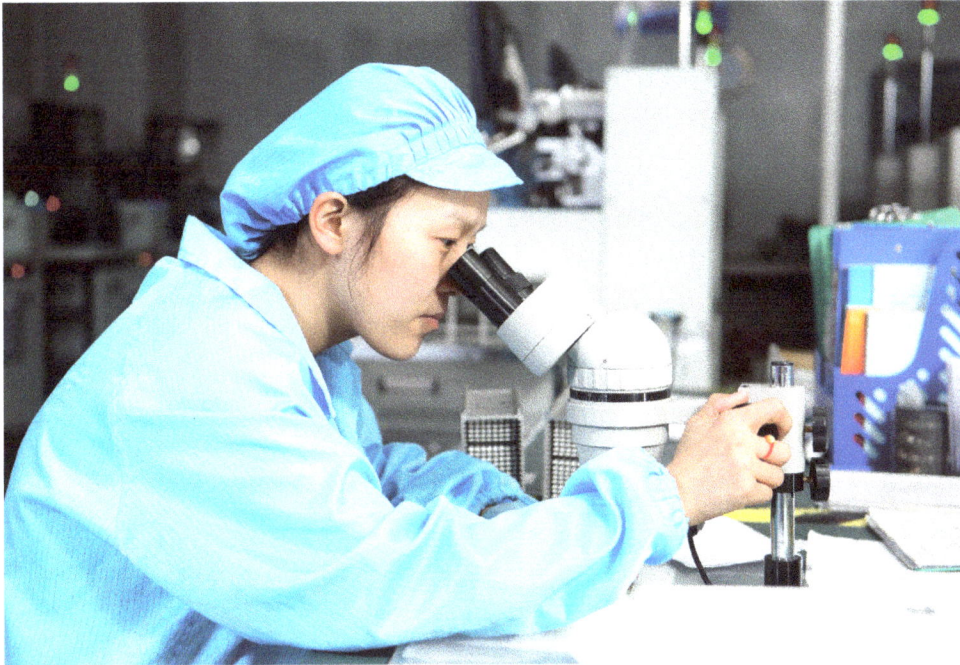

Image reproduced with permission from: ArtWell/ Shutterstock.com. All rights reserved.

THIS CHAPTER PROVIDES:

- A structure and quick checklist approach for gathering data in any workplace improvement project.
- The basics of task analysis so that the analyst gains a clear idea of the intended and/or current operations, in order to select a scope for improvement.
- Hierarchical task analysis (HTA).
- Tabular task analysis (TTA).

How to cite this book chapter:
Berlin, C and Adams C 2017 *Production Ergonomics: Designing Work Systems to Support Optimal Human Performance.* Pp. 127–138. London: Ubiquity Press. DOI: https://doi.org/10.5334/bbe.g. License: CC-BY 4.0

WHY DO I NEED TO KNOW THIS AS AN ENGINEER?

If the authors of this book were to characterize ideal engineering work in a single word, it would be: structured. In the study of production systems, we are constantly striving to balance a large number of considerations and parameters that we want to optimize at the same time. Without a structured way of doing this, it is quite possible to not only miss many opportunities for improvement, but also to overlook losses of efficiency and productivity that could have been fixed if they had only been identified in time. Sometimes, there are conflicting optimization goals in the same system, and without a structured way to identify them, there is a risk for technical design improvements that actually end up making other aspects worse.

To avoid this, the engineer needs a structured way to describe what is happening in the workplace; especially when observing human behaviour, it is helpful to have methods that can show the difference between *intended* work procedures, and how things are really done.

When we are able to break down a job into tasks, we can see (and describe) very clearly when things are going as they should, and where specific steps involve actions that we can improve. A good task breakdown clarifies two things: what we want to do and what we are actually doing.

WHICH ROLES BENEFIT FROM THIS KNOWLEDGE?

The *system performance improver* needs to collect data in a way that allows comparisons of before and after states. Also, for this role in particular, exact knowledge of how tasks are carried out – and *meant* to be carried out – is essential information in order to target improvement actions effectively. This chapter supplies ideas for the various ways that exist for collecting work environment and work data, and how to ethically and efficiently gather data from humans and about humans as they work.

Likewise, the *work environment/safety specialist* may be well served by a good grasp of task analysis, in order to make the process of data collection efficient. Many of the ergonomics evaluation methods presented in Chapter 8 require specific measurements and surveying of qualitative as well as quantitative data. From a time and effort perspective, knowing ahead of time what type of data is important to make a workplace assessment (using appropriate equipment) helps this role avoid "overcollecting" data and spending time analysing unnecessarily.

Before carrying out any workplace improvement project, it is crucial to have a clear and structured idea of what conditions make up the current state of the workplace. There are a lot of factors to take in when assessing a workplace; as evidenced in Part 1 of the book, the human capacity for good work performance is influenced by very many different inputs and internal responses in the body and mind. In order to not get overwhelmed or waste time wondering which aspects to take into account, workplace improvers may decide to limit their scope to just examining one or a few isolated aspects – but this is seldom the best approach, since optimizing just one aspect at a time can have unintended effects on other performance factors. When the work environment is to be analysed for the first time, or the first time in a long time, it pays to have an approach for a holistic assessment of the workplace. We will in later chapters look at ergonomics evaluation methods specifically, but additional gains and improvement ideas may come from looking at work environmental physical factors, psychosocial and teamwork aspects, and available cognitive support.

The following three phases are necessary to conduct a holistic assessment of workplace ergonomics:

• Data collection
• Task breakdown
• Ergonomic evaluation

While the results obtained from using methods provide a good indication of where changes should be made, ultimately these methods are only instruments, and the responsibility to draw conclusions, make decisions, changes and recommendations is down to you, the production engineer.

Having a basic knowledge of both the strengths and weaknesses of each method is key when deciding which one will be used to conduct an evaluation.

7.1. Data collection involving humans

To design effective and healthy workplaces, there is a need to understand the current state of the workplace and its associated tasks. The systematic gathering of this necessary information is known as data collection, which is done to answer a question – in our case, identifying work tasks with a high chance of causing injury. Knowing how to choose methods for the collection of data is the first step in conducting a successful and valuable study. Regardless of the type of data collection method being used there are a number of best practices that should be followed:

• Be structured.
• Be systematic.
• Be ethical and respectful towards your human participants.
• Be ready to handle (analyse) your data.
• Be ready to present what your data says and express how dependable it is.
• Truthfully present any limitations there may be to the relevance of your findings.

7.2. Data collection approaches

There are two main approaches towards data collection involving humans; a quantitative approach, which seeks to measure and quantify, or a qualitative approach, which seeks to understand processes, reasons and interdependencies. Deciding which one to use is very dependent on the nature of your research goal. However, it is counterproductive and wrong to automatically think that "qualitative" is synonymous with "subjective"[1], and that "quantitative" is synonymous with "objective"[2]. This is a misconception, but there is also an easy explanation to why it arises.

A qualitative approach is exploratory, answering questions of "how?" and "why?" (which is fully possible to do in an objective fashion). To a high degree, qualitative data collection involves interaction with people (e.g. interviewing and observing them), and is used for initial learning about previously unknown behaviours, defining new thought concepts, and recognizing trends and relationships between events. Typically this approach enables richer, detailed, in-depth answers to be obtained, but due to the time it takes to use qualitative data collection methods, such studies tend to involve small sample sizes/fewer people. If the aim is to learn about the nuances and variations of an unknown area of knowledge, a qualitative approach is suitable.

On the other hand, quantitative studies are suitable for examining relationships between previously well-described concepts and measurable changes in status. It is very important in quantitative method to be precise about what is being measured, so that there cannot be multiple interpretations of the results. This approach is more numbers-driven, as the aim is to measure and quantify, answering questions of "what?" or "how many?" As a result, answers to quantitative questions tend to be very brief, entirely avoiding explanations of why and how.

Quantitative methods have four main modes:

- **Census:** obtains data from every member of a population
- **Sample survey:** obtains data from a subset of a population (a sample), in order to estimate population attributes "well enough"
- **Experiment:** a controlled study in which the researcher attempts to understand cause-and-effect relationships by deliberately manipulating inputs and influencing factors
- **Observational study:** Like experiments, observational studies attempt to understand cause-and-effect relationships, but the researcher deliberately avoids manipulating any of the events

To obtain valuable quantitative results, the concept or theory being measured needs to have a scale by which it can be counted. Because the aim is to examine exact relationships between changes in quantity, the results are dependent on a large sample size (in the hundreds or thousands to be statistically significant), for the purpose of generating statistics that can indicate how relevant the theory is for the population being studied. However, statistics can also be carried out on the basis of data that is not objective in nature – opinion surveys are an example of this.

7.3. Carrying out a research study involving humans

When conducting data collection there are three key phases:

- Setting up the study
- Carrying out the actual data collection
- Analysing and presenting the data

In any study involving humans, a basic level of ethical standards is that of *informed consent,* i.e. all participants should be informed about the purpose of the study, what is expected of their involvement, and how any collected data will be handled afterwards. Participation should be voluntary (with consent given in documented form) regardless of the nature of the study (observation, experiment, interviews, etc.), so participants should also be informed of the option to say no and decline participation. The study should also ensure that no humans will be harmed or have their personal data compromised by any of the activities or how the data are handled post-collection. If a study is carried out in an organization (such as a university), there may be an ethics board who must screen the study plan and approve it before it is permitted to go on.

These phases can be broken down more specifically into 12 steps:

1. **Define the overall goal of the research** – Are we collecting data and analysing it in order to map/describe? To define? To visualize? To quantify?

2. **Determine the type of research question for the study** – Is the area well defined or undefined; should a quantitative or qualitative study be conducted?

3. **Determine the scope, time frame and sample** – What limitations exist? Time frame available and time each activity will take, geographical or cultural boundaries, etc. Will a random, purposive or convenience sample be used?

4. **Determine stop criteria** – When has enough data been collected?

5. **List ethical considerations** (usually regarding confidentiality) – Respect participant's privacy and inform them of exactly what you are going to do with their data. Depending on your organization's requirements, present the study design to an ethics board for approval.

6. **Choose your collection method and tools –**
   Methods:
   - *Observations:* An attempt to gain an unprovoked understanding of the task at hand while observing people's behaviour. In *think aloud* observations the participant talks out loud, explaining their reasoning and thought process while completing a task.
   - *Interviews/focus groups:* Suitable for in-depth qualitative studies with few participants, can be structured, semi-structured or unstructured. Try to avoid yes or no answers, which will stop the flow.
   - *Questionnaires/surveys:* Mainly used to quantify and measure different occurrences, which can be made into statistics. Answers are often presented on a scale. Some brief free-text responses may be possible to collect, but are sometimes difficult to interpret correctly due to lack of context.
   - *Case studies*: A specific context and event is studied in detail, to gain rich understanding.
   - *Document studies:* Existing documents (e.g. at companies) are studied and interpreted.
   Tools: camera, audio recording device, measuring equipment (tape measure, goniometer, weigh scales, etc.), stopwatch, checklist, estimation scales.

7. **Make a structure for your data collection** – How are you going to store and manage your data (e.g. by date, theme or participant)?

8. **Recruit or select participants** – This can be a time-consuming process, since participation must be voluntary. Participants must be informed of the purpose of the study and their part in it, and be informed of how the collected data will be handled. To boost participation, recruiting can often involve incentives (common examples include offering snacks and beverages or tokens of appreciation such as gift vouchers or movie tickets).

9. **Pilot test the data collection and tweak** – Use a small sample to test the method, and if necessary make modifications to the setup or questions before using a large sample size.
10. **Carry out the collection** – Ensure equipment is working and keep results well structured.
11. **Analyze the data** – "Let the data speak" and find meaningful trends, patterns or relationships within or between data sets.
12. **Present your findings** – Present findings together with limitations (such as sample size, time constraints, sample or environmental preconditions).

7.4. Task breakdowns

When studying how humans react to their work environment and the tasks they are required to carry out, utilizing a *task analysis* (Annett and Stanton, 2000) can provide a systematic working description of the task. Such an analysis provides accurate information of how tasks are performed in reality, which can be compared with how tasks should be performed and/or how tasks are perceived to be performed. In task analysis, a detailed description is established of the working task and all its necessary sub-tasks, showing how all the tasks are interconnected at various levels. This approach can be used to predict difficulties, evaluate performance and identify risks.

The analysis starts by establishing what the overall goal of the task is. Once we have broken down the overall task into sequential chunks, it is possible to isolate certain postures or motions that, if left unchanged, could contribute to long-term damage or MSDs. The necessary information to draw up a task analysis can come from interviews, observations, manuals or past experience. Task analysis has been extensively developed in research, particularly in connection with cognitive ergonomics and engineering, but we provide a short overview of the basics in this book hoping to encourage workplace designers to include evaluation of tasks in their coverage of finding improvement potentials.

7.5. Hierarchical Task Analysis (HTA)

A hierarchical task analysis (HTA; Annett, 2003) is task analysis method that provides an extensive description of all the necessary tasks to achieve a main goal in a hierarchal structure. Although terminology may vary somewhat, the following terms are normally used to identify different aspects of the task at various levels when conducting an HTA:

Glossary of HTA terminology

GOAL	External task resulting in a verifiable change of state, such as "making coffee" or "assembling a [product]"
TASKS	Activities necessary to achieve goals, sometimes using a device. Also called "sub-goals" (e.g. by Stanton, 2006)
SUBTASKS	Components of tasks
OPERATION	Simple task performed, lowest-level single action

DEVICE	Tool, machine or technique appropriate for achieving goals
PLAN	Number of tasks or actions linked into a sequence and describing rules and dynamics

These terms can be visualized in a hierarchical structure as shown in Figure 7.1.

In theory, it is possible to keep breaking down all the tasks until an exhaustive detailed list is established, right down to "sending nerve signals to muscle". However, such detail is rarely useful for engineering purposes, so it is necessary to determine the "stop criteria" for the degree of detail necessary. Knowing when to stop is just as much part of the process as identifying the subtasks is. It is okay to only expand the relevant tasks you are interested in and stop others at a higher level. Typically you stop decomposing when it becomes no longer relevant for the subsequent analysis. When visualizing a HTA, putting a solid line under the box shows that this is the deepest level of detail chosen for that specific task or operation.

How to carry out an HTA

A more formalized procedure for carrying out an HTA is described by Stanton (2006), in the form of the following steps (adapted from Stanton, 2006; p. 62–64):

1. **Define the purpose of the analysis**
   (e.g. system design, developing personnel specifications, analysis of workload, etc.)
2. **Define the boundaries of the system description**
   (i.e. which people and equipment will be considered in the analysis)
3. **Try to access a variety of sources of information about the system to be analysed**
   (This is in order to assure and validate the accuracy of the HTA; may include observation, interviews, expert consultation, manuals, simulation, etc.)

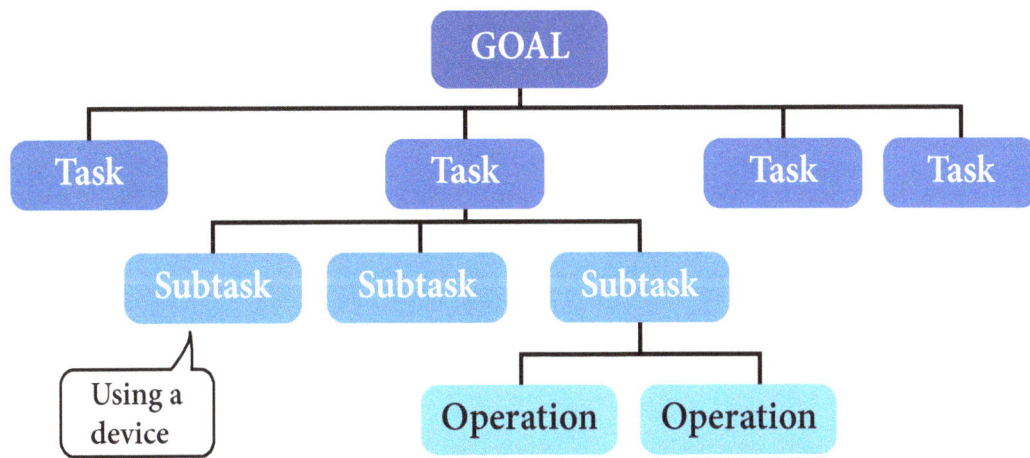

Figure 7.1: HTA structure.

Illustration by C. Berlin.

4. **Describe the system goals and sub-goals**
(What is to be done? To what performance standard? Under what conditions?)

5. **Try to keep the number of immediate sub-goals under any super-ordinate goal to a small number (aiming for 3–10)**
(Although there are exceptions, about 3–10 sub-goals are appropriate; if there are more, the analyst should consider whether some of these can be grouped into a super-ordinate goal to clarify the task overview.)

6. **Link goals to sub-goals, and describe the conditions under which sub-goals are triggered**
(This is where plans are formulated to guide the sequence and iterations between the specified sub-goals. Plans indicate which conditions trigger a sub-goal, when the purpose is fulfilled and the next step is to be taken.)

7. **Stop re-describing the sub-goals when you judge the analysis is fit for purpose**
(The level of description depends on the purpose of the analysis, so when to stop is up to the analyst – and is certainly easier if it is known who will use the information and how.)

8. **Try to verify the analysis with subject-matter experts**
(Verification with experts is important to make sure the analyst has interpreted the system goals and operations correctly, and can add the benefit of transferring ownership of the analysis to the experts)

9. **Be prepared to revise the analysis**
(As described in Stanton (2006 p. 64): "The number of revisions will depend on the time available and the extent of the analysis (…)". This may mean that several iterations are required to make the HTA accurate. Stanton (ibid.) continues: "It is useful to think of the analysis as a working document that only exists in the latest state of revision.")

HTA can be applied to a wide variety of work tasks, due to its flexibility in scope. Figure 7.2 shows an HTA where the overall goal is to unload a car.

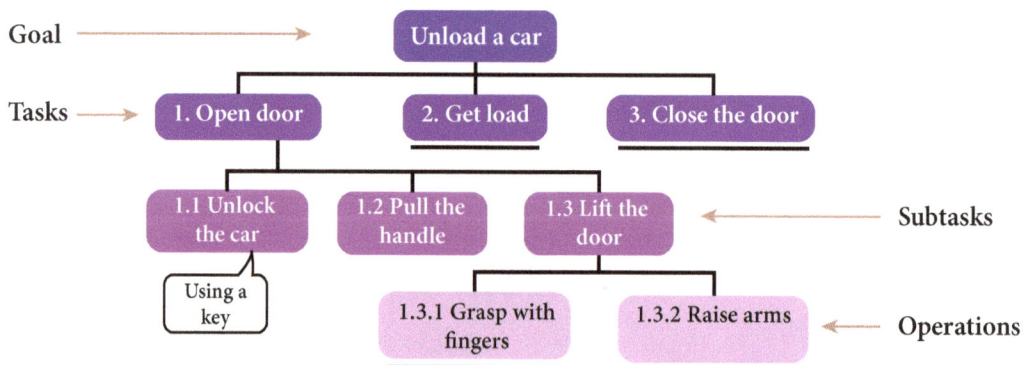

Figure 7.2: HTA Example – unloading a car.

Illustration by C. Berlin.

In this example, it can be seen that the analyst identified that the human needs to grasp the door handle with their fingers to open the door, but stopped going into further detail (note solid line under the operation box) about how many fingers will be used and which specific muscles will contract, as this is hardly necessary for understanding or analysing (or improving) this task.

7.6. Tabular task analysis (TTA)

It is also possible to visualise the same operations and tasks in a tabular form. The previous example of unloading a car could also be represented as in a table as shown in Table 7.1.

A task analysis can be an effective way to gain an overview of all the necessary tasks and operations involved with achieving the overall goal.

Once you have an idea of how tasks are intended to be (or are currently) carried out in a workplace, you have a basis on which to state your improvement goals and a way to focus and delimit your intervention efforts. In this way, you can be effective and efficient in your proposals, and clear in your communication about them towards other stakeholders.

Table 7.1: TTA Example – unload a car.

0	Unload a car
1	Open door
1.1	Unlock the door
1.2	Pull handle
1.3	Lift door
1.3.1	Grasp with fingers
1.3.2	Raise arms
2	Get load
3	Close door

Study questions

Warm-up:

Q7.1) When performing a study on humans, what are some basic ethical requirements on the work of an engineer or researcher?

Q7.2) For the purposes of workplace improvement, what is the difference between an observation and an experiment?

Q7.3)   Why are the following HTAs incorrect in principle?

a)

b)

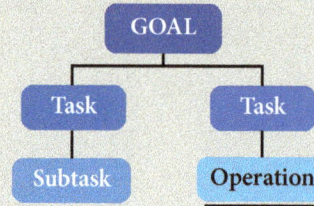

Look around you:

Q7.4)   Think of an everyday multiple-step task with a clear overall goal – like making a cup of tea or borrowing a book from the library – and try to perform a HTA on it. Can you determine the order in which steps are carried out, any necessary repetitions until the desired outcome is reached, and determine a lowest level of breakdowns (operations) for each step?

Connect this knowledge to an improvement project

- At the start of any workplace improvement project, it is crucial to understand the purpose and goals of what is done in that workplace. Alongside an observation and interview, a task analysis can help to form a basis for structured discussions about where and during what task risks are occurring; it also facilitates follow-up of whether interventions addressed the right target problems.
- The knowledge in this chapter can be used to plan time spent at worksite visits to investigate potential improvement potentials efficiently and comprehensively.
- The approaches described here allow for a separation of data collection (e.g. using recording devices) and data analysis, so that not all work needs to be done on-site where there is a risk of disrupting on-going work.

Connection to other topics in this book:

- Some ergonomics evaluation methods (Chapter 8) are task-oriented and it is sound practice to be able to identify the circumstance in which a particular posture, force or time exposure occurs. It is also easier (thanks to an awareness of overall goals for a task) to determine the reason for the risk – it may have to do with achieving a particular level of quality or performance speed.
- From an economical perspective (Chapter 11), it is wise to carry out a task breakdown to determine whether particularly crucial tasks that add great value (e.g. due to high demands of precision, quality and/or speed) are also associated with ergonomic pitfalls that risk being a chronic cause of unnecessary costs due to injuries, inefficiencies and scrap.

Summary

- A structured approach to understanding a workplace provides a dependable foundation for identifying and addressing improvement potentials.
- A work task can be broken down into elements to aid in analysis and identify risk areas, using a task analysis method.
- A number of methods are available for data collection and analysis, some of which are theoretically based while others actively involve workplace stakeholders.
- Using a combination of workplace observation, interviews and task analysis helps to give the workplace improver (and many other stakeholder roles) a good overview of what is meant to be done in a workplace, and at which points in the task-to-operations sequences there are ergonomic risks.
- Although the groundwork above may be considered time-consuming, it greatly facilitates discussions of where to direct intervention efforts and new design solutions, bringing them down to an appropriate level of precision. It also facilitates follow-up by pinpointing whether interventions and investments are appropriately targeted in scope to solve identified problems.

Notes

[1] Subjective data put the interviewee's personal perception, opinion and experience into focus; generally the answers are only possible for the asked person to verify (e.g. how they prioritize tasks, how much pain they are experiencing or how they perceive their workload).

[2] Objective data is possible to verify in an impersonal manner. This includes historical documentation, numbers, measurement from instruments, previously known facts, etc. (e.g. the temperature variation in a room over time or measured forces).

7.7. References

Annett, J., & Stanton, N. A. (Eds.). (2000). *Task analysis*. London: Taylor & Francis.

Annett, J. (2003). Hierarchical task analysis. *Handbook of cognitive task design*, 2: 17–35.

Stanton, N. A. (2006). Hierarchical task analysis: Developments, applications, and extensions. *Applied Ergonomics*, 37 (1):55–79.

CHAPTER 8

Ergonomics Evaluation Methods

THIS CHAPTER PROVIDES:

- Guidelines for conducting data collection and choosing a research approach.
- Heuristic evaluation.

How to cite this book chapter:
Berlin, C and Adams C 2017 *Production Ergonomics: Designing Work Systems to Support Optimal Human Performance.* Pp. 139–160. London: Ubiquity Press. DOI: https://doi.org/10.5334/bbe.h. License: CC-BY 4.0

• Various posture analysis methods; REBA, RULA and OWAS.
• Biomechanical assessment methods; NIOSH, Liberty Mutual materials handling tables.
• Combined methods; KIM.
• Standards and provisions from different countries.

WHY DO I NEED TO KNOW THIS AS AN ENGINEER?

Once we have a clear idea of what goes on in a workplace when workers perform tasks, we may be able to use our knowledge of what is beneficial or harmful to human work abilities to determine where the improvement potentials are. But analysing the risks and improvement potentials based on that knowledge may get complicated, easy to over- or understate, and difficult to communicate to other stakeholders who are not educated about the human body's needs and capabilities. In other words, a reliable shortcut is needed to help us decide what to target in improvement work – preferably in a way that simplifies and quantifies the risk levels to make comparisons easier.

In this chapter a number of analytical methods for assessing (mainly physical) ergonomics will be added to your "toolbox". This enables you to evaluate workplaces with respect to one or more of the physical loading factors mentioned in Chapter 3 (posture, forces and/or time), sometimes in combination with other work environment aspects such as those mentioned in Chapter 11. Most of the described ergonomics evaluation methods help you structure your analysis and prioritize which problems to target first by identifying the greatest risks for physical injury, ranking them in order of severity, and indicating which body segments are at risk. This structure of assessment makes it easier to communicate your decision basis to other stakeholders and justify particular interventions that target physical ergonomics root causes.

WHICH ROLES BENEFIT FROM THIS KNOWLEDGE?

Both the *system performance improver* and the *work environment/safety specialist* need knowledge of the methods presented here in order to communicate effectively with other stakeholders about physical loading risks that are present in the workplace. The eventual translation of risk into "severity levels" (often *red/yellow/green* classifications) is helpful in communications with management and other stakeholders tracking KPIs, but being able to arrive at these classifications requires solid knowledge of how to appropriately choose a method that captures the appropriate risk perspective. These roles may also find themselves communicating with medical or health and safety professionals with a more individual-risk focused perspective. The engineer with knowledge of these methods is given a platform for discussing how risk elements associated with particular body segment loading can be targeted.

The use of established, documented methods is important when conducting any assessment, as it ensures that analyses are conducted in a standardized, repeatable way. So should someone else carry out the same analysis at a later point in time, it is possible to fairly compare the results of both studies in a meaningful way.

A number of methods exist which enable us to study, analyse and evaluate humans while they are carrying out work tasks. Combining such methods with knowledge about the anatomical structure of the body and how it reacts to loading enables us to design effective and healthy workplaces. This chapter will introduce a number of useful methods and guidelines to evaluate whether humans are at risk when performing work tasks and interacting with their surrounding environment. Most methods also provide a guide for prioritization, helping the analyst determine which problems to address first.

8.1. Heuristic evaluation (HE)

One rough inspection or "checklist" method that enables general ergonomics issues to be identified is a *heuristic evaluation*. Heuristics can be explained as "rules of thumb" or "shortcuts" to decisions, based on conventional knowledge. With this method, a workplace or work tasks are evaluated according to a set of accepted principles, based on theoretical knowledge of human abilities and physical limitations, alongside past experience of how a design should be to work effectively. Deviations or causes for concern are noted and prioritized. Using a set of heuristics that have been predetermined before the study is known as a *structured* heuristic evaluation, but there is also some benefit from taking an *unstructured* approach and making up a list of heuristics as you go along during the evaluation. In the case of an unstructured approach, a high degree of theoretical knowledge is required on the part of the analyst, in order to conduct a meaningful and valuable study. For this reason, heuristic evaluations demand the participation of an expert to be accepted as reliable. Examples of common heuristics to consider when analysing a workstation are:

- No bending of the neck backwards.
- Pinching grasps should be avoided.
- Bending and twisting of the spine should be avoided.
- For heavier work, a working height of 100–250 mm below elbow height is recommended.
- For light work, a working height of 50–100 mm below elbow height is recommended.
- For push buttons, a height between elbow and shoulder is recommended.
- Lifting should be carried out close to the body.
- Adaptation to anthropometric variation (different body sizes) should be possible.

To conduct a heuristics evaluation the following procedure should be followed:

1. **Select heuristics to evaluate with**
   - Use existing (structured evaluation)
   - Create your own (structured evaluation)
   - Unstructured evaluation
2. **Evaluate the design based on the heuristics**
   - Note deviations from heuristics

- Explain why something is a problem (with respect to the heuristics) – simply identifying a problem is not enough
- Use task analysis as a base (for example, HTA)

3. **Assemble deviations and identify problems**
   If there is more than one evaluator this is done jointly, and a protocol created.

4. **The severity of the problems and deficiencies are assessed (if possible):**
   0 = Not an ergonomic problem.
   1 = Inconvenience problem; does not need to be fixed unless extra time is available.
   2 = Minor ergonomic problem; fixing should be given lower priority.
   3 = Major ergonomic problem; important to fix, high priority.
   4 = Very serious ergonomic problem; need to be fixed, high risk of injury.

5. **Reporting of results**
   - Compile into a protocol
   - Show result with task analysis

When showing the result on an HTA, it can be beneficial to use a colour-coded system to highlight the severity of the problems, hence indicating a priority order for design changes. For example, the HTA in Figure 8.1 shows areas for concern based on a heuristics evaluation of the task of changing tyres on a car.

To ensure this method is carried out effectively, it is important to be aware of both its strengths and weaknesses. While it is a fast, resource-efficient method that is simple to carry out, it is also limited in value due to its subjective nature, limited scope and somewhat unsystematic evaluation approach.

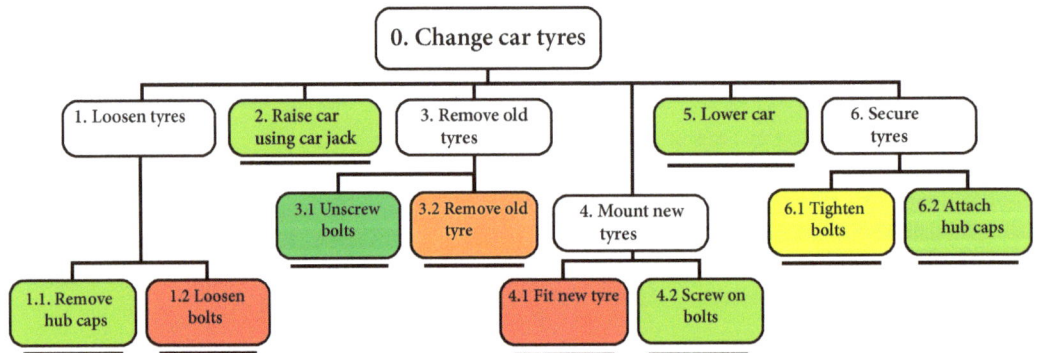

Figure 8.1: HTA demonstrating severity of issues identified during HE, when changing car tyres.
Illustration by C. Berlin.

Where the colour levels indicate:
0 = No Problem
1 = Inconvenience
2 = Minor ergonomic problem
3 = Major ergonomic problem
4 = Very serious ergonomic problem

8.2. Methods for evaluating physical loading

A number of different established methods exist for assessing physical load. These methods fall under three broad categories:

- Posture-based analysis
- Biomechanics-based analysis
- Analysis based on a combination of environmental factors

Table 8.1 provides a summary of some of these methods, describing their main function and which category they belong to. There are many, many more ergonomics evaluation methods available, but this list aims to present a variety across the categories mentioned. To give detailed instructions for each method would make this book very cumbersome, and it is often best to use the source materials for this purpose, so each method description in this chapter also provides links to instruction sheets for each method, as made publicly available.

8.3. Posture-based analysis

Having provided an anatomical basis for understanding the capabilities of the human body in earlier chapters, we will now go onto discuss posture-based methods for studying work tasks from a physical ergonomics perspective. Posture-based ergonomics evaluation methods use point-based systems to rank identified areas of concern. Typically, the more the body deviates from the neutral standing position, the worse the working posture is, thus resulting in a higher score. The selected methods presented here are quick and simple to conduct and are based purely on observations, making them somewhat vulnerable to interpretation. By looking at different regions of the body and joint angles, the loads experienced by the body are ranked on a pre-determined scale of risk severity.

Generally, posture-based observation methods are *screening tools,* meant to give a risk estimation for system designers to prioritize which risk factors to address first; generally the point is to eliminate causes of high rating points as a first step. If the screenings return results indicating some uncertainty as to the risk level, more in-depth analyses may be recommended, perhaps using a different risk assessment tool. While these methods are quick and simple ways to evaluate posture, they are somewhat limited as they don't always consider time exposure or accumulating loads, and are subjective due to the element of observation. Generally, the same method should be used before and after a design change to monitor the impacts of design changes, to see if sufficient posture improvements have been made according to the same set of posture assessment criteria.

RULA (Rapid Upper Limb Assessment)

RULA and REBA are two similar methods that can be used to quickly screen and identify harmful postures. RULA (McAtamney & Corlett, 1993) is more suited to hand-arm intensive work, having been developed to study sitting assembly work in textile confectionery industry, while REBA (Hignett & McAtamney, 2000) covers whole-body intensive work, as it was developed in a hospital/healthcare context. Both methods focus on one specific posture that occurs during the work tasks. This posture

Table 8.1: Summary of physical load assessment methods and associated resources for manuals, worksheets etc.

Method	Purpose and online resources for manuals and worksheets	Postural analysis	Materials handling tasks	Additional aspects (time, space, intensity, speed, etc.)
RULA (Rapid Upper Limb Assessment)	Upper body & limb assessment, screening of postures **Introduction, paper- and spreadsheet-based worksheets:** http://ergo.human.cornell.edu/ahRULA.html	X		
REBA (Rapid Entire Body Assessment)	Whole-body posture analysis, screening of postures **Introduction, paper- and spreadsheet-based worksheets:** http://ergo.human.cornell.edu/ahREBA.html	X		
OWAS (Ovako Working Posture Analysing System)	Whole body posture analysis, screening of postures over time **Manual:** Available as a handbook/training publication, see Reference list (Louhevaara and Suurnäkki, 1992)	X		X
NIOSH Lifting Equation	To identify whether a lifting load is acceptable for workers **Applications Manual for the Revised NIOSH Lifting Equation including tables of multipliers:** http://www.cdc.gov/niosh/docs/94-110/ and http://www.ergonomics.com.au/use-revised-niosh-equation/		X	
Liberty Mutual manual materials handling tables	To identify the portion of a specified male or female population that should be able to lift, lower, carry, push or pull as part of their daily work, without risks for MSDs **Tables:** https://libertymmhtables.libertymutual.com/CM_LMTablesWeb/taskSelection.do?action=initTaskSelection		X	

Method	Description			
JSI (Job Strain Index)	Assessment of risk for upper extremity disorders, particularly focused on repetitive tasks **Method description and worksheets:** http://ergo.human.cornell.edu/ahJSI.html and http://www.theergonomicscenter.com/graphics/ErgoAnalysis%20Software/Strain%20Index.pdf	×		×
KIM (Key Indicator Method)	To identify level of risk associated with manual material handling tasks – exists in three specific forms for different materials handling cases **Lifting – Holding – Carrying:** http://www.beswic.be/en/topics/msds/slic/handlingloads/29.htm **Pushing – Pulling:** http://www.beswic.be/en/topics/msds/slic/handlingloads/19.htm/30.htm **Manual Handling Operations:** http://www.baua.de/de/Themen-von-A-Z/Physische-Belastung/pdf/KIM-manual-handling-2.pdf?__blob=publicationFile&v=4	×	×	
EAWS (Ergonomic Assessment Worksheet)	Screening worksheet developed by the MTM (methods-time measurement) community to evaluate ergonomic risk exposure aligned with predetermined time standards **Method description and worksheets:** http://mtm-international.org/risk-screening-the-ergonomic-assessment-work-sheet-eaws/	×	×	×
RAMP (Risk Assessment and Management tool for manual handling Proactively)	*RAMP I* **Checklist for screening physical risks for manual handling** https://www.ramp.proj.kth.se/ *RAMP II* **In-depth analysis for assessment of physical risks for manual handling** https://www.ramp.proj.kth.se/	x	×	×
HARM (Hand Arm Risk-assessment Method)	Hand- and arm-focused analysis method to identify and screen repetitive tasks **Manual and worksheets:** https://www.fysiekebelastingbeoordelen.tno.nl/download/HARM_Manual_paper-based_harm.pdf	×		×

is generally identified through observations and discussions with the worker. Generally, postures that occur frequently, last for a prolonged period of time, involve large forces or muscular activity, cause discomfort, or are considered to be extreme are the ones typically selected for analysis. During an assessment, the whole task is observed and key postures of interest are identified. These data points can then be captured visually (e.g. filmed, photographed or observed) enabling a RULA score to be calculated using the RULA assessment form. For conditions that are considered to worsen the posture, additional "penalty" points are added. The final score is used as an indication to show how soon it is necessary to do something about the observed posture. In a RULA analysis, the positions of six different body regions are considered: upper arm, forearm, wrists, neck, trunk (upper torso) and legs. Based on the deviations of each body part from the "neutral" position, the weight of any loads, and the nature of movements (static or dynamic), an overall score is calculated. This final score between 1–7 corresponds to a ranking, which indicates to the analyst whether the posture presents an injury risk. It is possible to conduct a RULA analysis within simulation software (this is discussed in more detail in Chapter 9); despite being older than REBA, RULA is more commonly found as an evaluation tool in simulation software.

Worksheets for paper-based RULA evaluations (in metric measures) are available at the link in Table 8.2.

REBA (Rapid Entire Body Assessment)

REBA (Hignett & McAtamney, 2000) is a similar method for evaluating body postures during work tasks, but unlike RULA it focuses on whole-body intensive work. Similarly to RULA, one specific posture that occurs during the work task is analysed to provide an overall score. A REBA analysis considers the same six body regions as RULA, but it goes one step further by also taking couplings and grips into consideration. Points are added for conditions that worsen the nature of the posture, and points can also be subtracted if something contributes towards lessening the loading impact of the posture (such as gravity-assisted postures). The final score between 1–15 is calculated using the REBA assessment form.

Worksheets for paper-based REBA evaluations (in metric measures) are available at the link in Table 8.2.

KEY CONSIDERATIONS FOR RULA AND REBA ANALYSIS

When conducting RULA and REBA the following points should be kept in mind:

- Is the posture caused by the environment (workplace) or materials being handled?
- Does the selected posture affect both tall and short workers?
- Did you assume that this posture is transient (a changing movement)?
- How often does this posture occur?
- What kind of strength does the position require?
- Would training help in eliminating the posture?

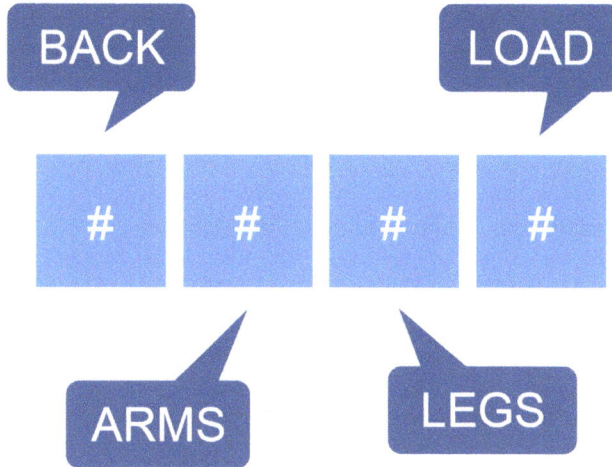

Figure 8.2: OWAS Score – each digit represents a posture or load assessment. Illustration by C. Berlin.

OWAS (Ovako Working Posture Analysing System)

OWAS, short for Ovako Working Posture Analysing System, is somewhat similar to REBA and RULA in that it provides a figure indicating how harmful a posture is (Louhevaara and Suurnäkki, 1992). Since it originated in the steel industry, the method was initially designed with heavy lifting in mind. The analysis result is a four-digit score describing posture (Figure 8.2), where the first value is concerned with the back, the second the arms, the third the legs and the fourth weight/external load. The end result highlights the areas where most of the riskiest work postures appear. The complete process necessary to carry out an OWAS analysis is described in Louhevaara and Suurnäkki (1992).

HARM (Hand Arm Risk-assessment Method)

HARM (Douwes and de Kraker, 2014) is a method developed by researchers at the Dutch institute TNO (the Netherlands Organisation for Applied Scientific Research), specifically tailored to analysing risks for MSDs in the hand and arm, and it takes into account both posture of the arms, wrists, neck and head, and also time aspects (including repetitiveness) and forces. The method exists as HARM1.0 (Douwes and de Kraker, 2014) and the updated HARM 2.0 (TNO, 2012) with reduces the relative weight of task duration, simplifies the force categories and includes some clarifications and changes to the instructions and the manual.

8.4. Biomechanics-based analysis

Ergonomic evaluation methods that utilise biomechanical calculations also exist. These methods tend to be based on the evaluation of work tasks that involve moving a load from one place to another by pushing, pulling, carrying, lowering or lifting it. Compared to observational posture-based analysis methods, they take longer to carry out and provide a strictly defined, more numerical result.

Liberty Mutual manual materials handling tables

Based on the initial research work presented by Dr Stover Snook and Dr Vincent Ciriello initiated in 1978 on materials handling, Liberty Mutual (an American insurance company) established an analysis tool to assess lifting, lowering, pushing, pulling and carrying tasks in the workplace (Snook & Ciriello, 1991). Given the costs associated with back disabilities and reduced productivity resulting from manual materials handling tasks, the tables provide criteria levels at which lifting can be judged as suitable or unsuitable for a well-defined working population. It is considered an objective risk assessment, in terms of being statistically backed. Since it is based originally on the work of Dr Snook, this method is also sometime referred to as "Snook's Lifting Recommendation" or "The Snook Tables".

A number of different tables provide information about both the male and female population, their capabilities for lifting, lowering, pushing, pulling and carrying. The tables can be used to identify the portion of the population that should be able to conduct such tasks as part of their daily work. The relevant table for the population and task at hand is selected, and the resulting maximum criteria value provides aid in modifying or redesigning the work task, to reduce or eliminate injury risk.

This method takes into consideration the vertical height of the item to be lifted, its weight, hand distance, hand height before and after the object has been lifted, frequency of tasks, and the distance it should be pulled, pushed or carried.

The updated tables and materials are accessible from the link in Table 8.2.

NIOSH lifting equation

This method, based on work conducted at the National Institute of Occupational Safety and Health (NIOSH) in America, is used to calculate whether lifting a load is acceptable (Waters et al, 1993; see Figure 8.3), using an equation which considers:

- Horizontal distance of the load from the worker
- Vertical height of the lift
- Vertical displacement during the lift
- Angle of symmetry between the mid-plane of the body and the direction of lift
- Frequency, duration of lifting
- Coupling between the worker's hand and the object.

By using a load constant of 23 kg, or 50 lbs (considered the maximum lifting weight permissible even under the best possible lifting circumstances) multiplied by factors that are ≤ 1, it is possible to calculate the RWL (Recommended Weight Limit) that can be handled by the majority of healthy people[1] during the working day:

$$RWL = LC \cdot HM \cdot VM \cdot DM \cdot AM \cdot FM \cdot CM$$

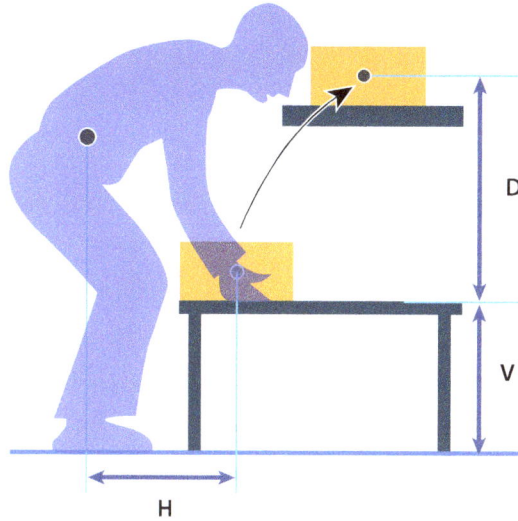

Figure 8.3: NIOSH Equation Schematic.

Illustration by C. Berlin, based on Bohgard (2009).

The Lifting Index (LI) is a related indicator that is calculated as follows:

$$LI = L/RWL$$

...where LI > 1 indicates an increased injury risk.

RWL = Recommended Lifting Weight
LI = Lifting Index
LC = Load Constant = 23 kg
HM = Horizontal Multiplier
VM = Vertical Multiplier
DM = Distance Multiplier
AM = Asymmetric Multiplier
FM = Frequency Multiplier
CM = Coupling Multiplier
L = Load Weight (the proposed weight)

Each of the multipliers is a decimal between 0 and 1, which decrease the LC when multiplied with it. These multiples are fetched from tables in the appendix of the manual for the revised NIOSH lifting equation (Waters et al., 1994).

It is important to note that there are a number of instances when the NIOSH lifting equation should *not* be used:

• When lifting with one hand
• When lifting work occurs for longer than an 8-hour shift
• When kneeling or sitting

- In a cramped space
- When lifting unstable objects (liquid containers, half-full boxes, etc.)
- When simultaneously carrying, pulling and pushing
- When using a wheelbarrow or shovel
- For quick lifting (high acceleration)
- On slippery floors
- In unfavourable environmental conditions, such as below 19°C, over 26°C or high humidity

The Applications Manual for the Revised NIOSH Lifting Equation including tables of multipliers are available at the link in Table 8.2.

8.5. Multi-aspect methods

JSI (Job Strain Index)

The Job Strain Index is another method used to identify injury risks during work tasks, but it is specifically focused on the upper extremities (wrist and hands) and is particularly beneficial when analysing repetitive jobs (Moore and Garg, 1995). This method takes in account the following aspects:

- Intensity of the exertion (IE)
- Duration of the exertion (DE)
- Efforts per minute (EM)
- Posture (HWP)
- Speed of work (SW)
- Duration of task per day (DD)

Each of the six factors are weighted based on tables using biomechanical, physiological, epidemiological and psychological criteria, and a final score is achieved by multiplying all the factors together:

$$JSI = IE \times DE \times EM \times HWP \times SW \times DD$$

The resulting score indicates the risk of developing a distal upper extremity disorder. It should also be noted that this method has a degree of subjectivity, as not all the factors can be explicitly measured. This method does not consider tasks involving vibrations or contact stress, which will obviously have a significant impact on the worker over time.

The method description and worksheets are available at the links in Table 8.2.

KIM (Key Indicator Method)

This analysis method was developed by the German Federal Institute for Occupational Safety and Health (2012) and is a screening method targeted at the manual handling of loads. There are three different variants of KIM: one for analysing work tasks and activities involving manual handling operations (MHO), another one for pulling and pushing (PP), and a third for lifting, holding and carrying (LHC). A series of rating points for a number of attributes including time, load, posture and working conditions (including work environment) are used to determine an overall score, which can then be checked against an established scale to determine the severity of the risk presented to the worker. A final score is achieved by adding the load, posture and working conditions ratings and multiplying the sum by the time rating. Rating points for each attribute are then determined by observing the task and selecting the most applicable characteristic from a series of predetermined tables.

The work templates for the three variants of the KIM method are available at the links in Table 8.2.

EAWS (Ergonomic Assessment Worksheet)

The Ergonomic Assessment Worksheet, EAWS, is a quick screening tool developed by the International MTM Directorate (IMD, 2015), an international interest organization for Predetermined Time Systems (which have a historically significant presence in the industrial engineering discipline). EAWS covers four risk areas: body postures, action forces, manual materials handling and upper limbs (with focus on high frequency). In keeping with the MTM emphasis on standardization, the method's acceptability criteria are aligned with several international standards, including CEN and ISO. The worksheet output is a green-yellow-red acceptability rating, based on a cumulative point scale.

The method description and worksheets are available at the links in Table 8.2.

RAMP (Risk Assessment and Management tool for manual handling Proactively)

RAMP is an observation-based method developed at Sweden's KTH Royal Institute of Technology for analysing workplaces for risks of MSDs (Lind et al., 2014; Lind, 2015; Rose, 2014). The method exists in the form of a simplified checklist for initial screening called RAMP I, where the analyst answers Yes or No to the occurrence of a number of risk types (covering the areas of postures, repetitive movements, lifting, pushing/pulling, influencing factors, physical strain and perceived discomfort), or as RAMP II, a refined analysis module to be used when the RAMP I analysis identifies risks that are uncertain in their cause or severity and require further analysis. The output from RAMP I is a colour-scale rating of green (low risk), grey (investigate further) and red (high risk). The output from RAMP II is a colour-scale rating of green (low risk), yellow (risk) and red (high risk), along with a sum of scores to help determine the prioritization of what to address first. Due to the inclusion of perceived operator discomfort, it is necessary to have an experienced operator (or more, to include variations in their work) to observe and talk to when analysing the task.

8.6. Standards, legal provisions and guidelines

Ergonomists and work designers in many countries use standards, guidelines and legal provisions to ensure that a workplace does not harm the workforce – sometimes these guideline documents have a powerful impact on achieving implementation of good workplace standards, as the legal status and recognition of the guidelines may be the only thing that will convince the management to take action to benefit the workers' well-being. Some countries have a strong tradition and established institutions continually release and update workplace guidelines that regulate the responsibilities of organizations and employees to provide and maintain a safe and healthy workplace. Table 8.3 summarizes a selection of national and international documents that guide and regulate the design of

Table 8.3: Examples of national and international/provincial codes, standards and guidelines primarily aimed at preventing work-related MSDs. Collected by the ILO and IEA for different countries (taken from Niu, 2010 p. 750).

Country	Document
Australia	• National Code of Practice for Manual Handling [NOHCS: 2005(1990)] • National Code of Practice for the Prevention of Occupational Overuse Syndrome [NOHSC: 2013 (1994)] • Manual Tasks Advisory Standard 2000 – Queensland • Code of Practice for Manual Handling 2000 – Victoria
China	• Law on Prevention and Control of Occupational Diseases (Article 13 of Chapter II Preliminary Prevention). 2002 • Occupational exposure limits for hand-transmitted vibration in the workplace (GBZ 2.2-2007), Measurement methods (GBZ/T 189.9), and Diagnostic criteria of occupational hand-arm vibration disease (GBZ 7) • Hygienic Standards for the Design of Industrial Enterprises (GBZ1) on workplace lighting and illumination • Guidelines for occupational hazards prevention and control (GBZ/T 211-2008)
European Community	• Directive 89/391 Introduction of measures to encourage improvements in the safety and health of workers at work • Directive 90/269/EEC Minimum health and safety requirements for the manual handling of loads where there is a risk particularly of back injuries to workers • Directive 2002/44/EC Minimum health and safety requirements regarding the exposure of workers to the risks arising from physical agents (vibration).
ISO	• ISO 11228-1 Ergonomics – Manual Handing – Part 1: Lifting and Carrying • ISO 11226 Ergonomics – Evaluation of static working postures • ISO/FDIS 6385:2003 Ergonomic Principles in the Design of Work Systems
Japan	• Guidelines on the prevention of lumbago in the workplace (1994).
Netherlands	• Working Conditions Act 1998
New Zealand	• Code of Practice for Manual Handling • Approved Code of Practice for the Use of Visual Display Units in the Place of Work • Occupational Overuse Syndrome (OOS) – Guidelines for prevention and management (1991) and Occupational Overuse Syndrome. Checklists for the evaluation of work (1991)
Norway	• Act Relating to Worker Protection and Working Environment (2003)
South Africa	• Occupational Health and Safety Act 1993
Spain	• Royal Decree 487/1997 Minimum health and safety provision relating to manual load handling involving risks for workers, particularly to the dorsolumbar region and the associated technical guide for the evaluation and prevention of risks associated with manual load handling. • Royal decree 488/1997 Minimum health and safety dispositions relating to work with equipment fitted with visual display units and the associated technical guide for the evaluation and prevention of risks associated with the use of equipment with visual display units.

Country	Document
Sweden	• AFS 2001:1 – Provisions of the Swedish Work Environment Authority on Systematic Work Environment Management, together with General Recommendations on the Implementation of the Provisions. • AFS 1998:1 – Provisions of the Swedish National Board of Occupational Safety and Health on Ergonomics for the Prevention of Musculoskeletal Disorders, together with the Board's General Recommendations on the Implementation of the Provisions
UK	• The Manual Handling Operations Regulations 1992 • The Health and Safety (Display Screen Equipment) Regulations 1992. • Upper limb disorders in the workplace. HSE, 2002 • Aching arms (or RSI) in small businesses, HSE, 2003 • Manual Handling Assessment Charts. HSE, 2003
USA	• OSHA, 2003: Ergonomics for the Prevention of Musculoskeletal Disorders. Guidelines for Poultry Processing. • NIOSH: Simple Solutions: Ergonomics For Farm Workers, 2001 • California Dept of Industrial Relations, 1999: Easy Ergonomics. A Practical Approach for Improving the Workplace • California Dept of Industrial Relations, 2000: Fitting the Task to the Person: Ergonomics for Very Small Businesses • State of Washington, Dept of Labor: WAC 296-62-051. Ergonomics • State of Washington, Dept of Labor: Fitting the Job to the Worker: An Ergonomics Program Guideline

work environments, with a main goal to prevent MSDs. The list was compiled in 2010 (Niu, 2010) as a collaboration between the International Labour Organization (ILO) and the International Ergonomics Association (IEA), and is shown as an overview – however, it is important to follow the updates of governing bodies for workplace health and safety, since they continually update requirements and guidelines. As an example, this book takes a closer look at the Swedish Work Environment Authority's most recent MSD-focused legal provisions.

8.7. Example: Swedish AFS provisions

The Swedish Work Environment Authority (*Arbetsmiljöverket*), formerly known as the Swedish National Board of Occupational Safety and Health (*Arbetarskyddstyrelsen*), is a legal entity in Sweden that works continually with releasing, renewing, amending and combining the legal guidelines, called *provisions*, for designing safe and healthy workplaces. All of these are enactments and updates of the Work Environment Act, which was established in 1977. The provisions are part of the Statute Book (AFS, *Arbetarskyddstyrelsens FörfattningsSamling*) and cover a very wide range of specific guidelines, ranging from physical loading and materials handling to chemical hazard restrictions, sector-specific guidelines, exposure limit values, psychosocial work environment and how to carry out systematic

Table 8.4: A succession of legal provisions from the Swedish Work Environment Authority that regulate the workplace in order to prevent work-related MSDs and other risks to worker well-being and safety.

AFS 1998:1	*Ergonomics for the Prevention of Musculoskeletal Disorders, provisions* (in English) (Swedish Work Environment Authority, 1998)
AFS 2001:1	*Systematic Work Environment Management (AFS 2001:1Eng), provisions* (in English) https://www.av.se/globalassets/filer/publikationer/foreskrifter/engelska/systematic-work-environment-management-provisions-afs2001-1.pdf
AFS 2009:2	*Workplace Design (AFS 2009:2Eng), provisions* (in English) https://www.av.se/globalassets/filer/publikationer/foreskrifter/engelska/workplace-design-provisions-afs2009-2.pdf
AFS 2012:2	*Physical ergonomics and work environment* (most recent, in Swedish) – *replaces AFS 1998:1* https://www.av.se/globalassets/filer/publikationer/foreskrifter/belastningsergonomi-foreskrifter-afs2012-2.pdf
AFS 2015:4	*Organisational and social work environment (AFS 2015:4Eng), provisions* (in English) https://www.av.se/globalassets/filer/publikationer/foreskrifter/engelska/organisational-and-social-work-environment-afs2015-4.pdf

work environment improvements. Each provision is marked by the year of publication and serial number, e.g. AFS 2012:2. Arbetsmiljöverket also assigns specific responsibilities to employers and employees to jointly carry the responsibility for workplace safety and health, although most of the specifics of workplace design befall the responsibility of the employer.

As described in Berlin et al (2009 pp. 941–942), "It is stated explicitly in the [AFS 1998:1] provision that an employer is responsible for continually maintaining a healthy workplace for the employees. The provision contains guidelines for assessment of work posture, duration of work cycles, lifting requirements and relevant conditions which increase or decrease the harmfulness of the work posture (e.g. duration of postures, repetitiveness, spatial dimensions of the workplace, weight of handled objects and possibilities of gripping them, freedom to autonomously decide when to take breaks, etc.) The values for boundary conditions in AFS-98 are stated to be valid for work shifts of four to eight hours in duration".

The Swedish legal provision AFS 2012:2 (Swedish Work Environment Authority, 2012) was released as an update of the previously used provision AFS 1998:1 (Swedish Work Environment Authority, 1998), and therefore the guidelines of both documents have more or less the same coverage. The guideline provides a variety of evaluation criteria for assessing the physical ergonomics of workplaces. The general principle is that most criteria are evaluated on a scale of green-yellow-red, with green being acceptable and red being unacceptable, and yellow requiring further investigation. The provision is intentionally vague with some room for interpretation, in order to be relevant for a variety of workplace types. This means that ergonomics expertise is recommended in order to use the guideline correctly (preferred analysts are physiotherapists or ergonomists), so most places that adhere to the AFS 2012:2 have in-house ergonomics specialists (such as an Occupational Health Service) to make the assessments.

Arbetsställning		Rött	Gult	Grönt
Sittande		Något av nedanstående förekommer **under en väsentlig del** av arbetsskiftet.	Något av nedanstående förekommer **periodvis** under arbetsskiftet.	Nedanstående gäller för **en väsentlig del** av arbetsskiftet.
	Nacke	– böjd – vriden – samtidigt böjd och vriden – kraftigt inskränkt rörelsefrihet	– böjd – vriden – samtidigt böjd och vriden – kraftigt inskränkt rörelsefrihet	– i mittställning – möjlighet till fria rörelser
	Rygg	– böjd – vriden – samtidigt böjd och vriden – kraftigt inskränkt rörelsefrihet – stöd för ryggen saknas	– böjd – vriden – samtidigt böjd och vriden – kraftigt inskränkt rörelsefrihet	– möjligheter till fria rörelser – väl utformat ryggstöd – möjlighet att växla till stående
	Axel/ arm	– handen i eller över skulderhöjd – handen utanför underarmsavstånd utan avlastning	– handen i eller över skulderhöjd – handen utanför underarmsavstånd utan avlastning	– arbetshöjd och räckområde anpassade till arbetsuppgift och individ – god armavlastning
	Ben	– otillräcklig plats för benen – inget stöd för fötterna – kraftigt inskränkt rörelsefrihet – ben- eller fotmanövrerat pedalarbete*)	– otillräcklig plats för benen – inget stöd för fötterna – kraftigt inskränkt rörelsefrihet – ben- eller fotmanövrerat pedalarbete*)	– fritt benutrymme – bra fotstöd – sällan ben- eller fotmanövrerat pedalarbete*) – möjlighet att växla till stående
Stående/gående	Nacke	– böjd – vriden – samtidigt böjd och vriden – kraftigt inskränkt rörelsefrihet	– böjd – vriden – samtidigt böjd och vriden – kraftigt inskränkt rörelsefrihet	– upprätt ställning – möjlighet till fria rörelser
	Rygg	– böjd – vriden – samtidigt böjd och vriden – kraftigt inskränkt rörelsefrihet – ostabilt eller lutande underlag	– böjd – vriden – samtidigt böjd och vriden – kraftigt inskränkt rörelsefrihet – ostabilt eller lutande underlag	– upprätt ställning – möjlighet till fria rörelser – möjlighet att växla till sittande
	Axel/ arm	– handen i eller över skulderhöjd – handen i eller under knähöjd – handen utanför ¾ armavstånd från kroppen	– handen i eller över skulderhöjd – handen i eller under knähöjd – handen utanför ¾ armavstånd från kroppen	– arbetshöjd och räckområde anpassande till arbetsuppgift och individ
	Ben	– otillräcklig plats för ben och fötter – ostabilt underlag – lutande underlag – ben- eller fotmanövrerat pedalarbete*)	– otillräcklig plats för ben och fötter – ostabilt underlag – lutande underlag – ben- eller fotmanövrerat pedalarbete*)	– fri rörelsemöjlighet på stabilt, halksäkert, jämnt och vågrätt underlag – inget ben- och sällan fotmanövrerat pedalarbete*) – möjlighet att växla till sittande

*) Benmanövrerat pedalarbete = bromsen eller kopplingen på en bil
   Fotmanövrerat pedalarbete = gaspedalen på en bil

Figure 8.4: Example of red-yellow-green guidelines from the AFS 2012:2 provision (Swedish Work Environment Authority, 2012 p. 37), showing work conditions at different risk levels for sitting and standing work.

Image reproduced with permission from: the Swedish Work Environment Authority. All rights reserved.

SELECTING A SUITABLE EVALUATION METHOD

In order to determine which method is best to analyse the task in question, the following questions should be answered to help you choose which method is most suitable.

What is the main characteristic of the task?

- Does the task involve hand-arm intensive work? Does it involve lifting, lowering, pushing, pulling or carrying? Is it a heavy, intensive task, or a light but constant load?
- Some tasks involve large forces, times or postures. Is one of these aspects dominant over the others?
  - Is the objective to describe, brainstorm or rate the work task?
  - Do we want a quick "screening" for a prioritization?

Nature of the problem

- Where do we predict that problems of incorrect working use will arise?
- Is the problem caused by motion or static postures?
- Is the task particularly intensive for a certain part of the body?

What can we measure in this task?

- Measure joint angles, time for the task, the forces or weights involved, and the distances travelled (if applicable). Are any of these measurements remarkable?
- If you find an extreme measurement, this might help you select an analysis method.

How does the task relate to the measurements of the person doing the work?

- Observe the person performing the task. Are there any specific operations of the task that increase the load, posture or discomfort because of the worker's body dimensions?

Study questions

Warm-up:

Q8.1)   What do the acronyms RULA and REBA stand for, and which work sectors did they originate from?

Q8.2)   What are the limitations of posture-based ergonomics evaluation methods?

Q8.3)   When designing lifting tasks, what limitation in applicability does the NIOSH lifting equation have for a female population?

Q8.4)   What are the limitations of heuristic evaluation?

Look around you:

Q8.5) Select and examine one (or more) of the workplace ergonomics standards listed in Table 8.3 or 8.4. Do they give a high level of detailed direction for how to design a workplace, or are they flexible in their criteria in order to suit many different work sectors?

Q8.6) Use one (or more) of the posture evaluation worksheets listed in Table 8.1 and try to recreate a posture that corresponds to the worst possible posture score in all categories. Is this a likely work posture for any reason? Try adjusting just one of the posture components to the lowest possible score. What is the impact on the total posture score?

Q8.7) Look through the list of ergonomics evaluation methods in Table 8.1 – can you imagine any particular work sector where the method would be suitable, based on the body segments and additional factors targeted in each method?

Connect this knowledge to an improvement project

- Established ergonomics evaluation methods can aid workplace designers in identifying and ranking ergonomic risks, so that the most hazardous risks are addressed as a first priority.
- Use the same method to evaluate risk before and after an intervention, to give a relative quantification of whether the risk level of a task has improved.

Connection to other topics in this book:

- Most ergonomics evaluation methods rely on the anatomical limits and principles described in Chapter 2 and Chapter 3.
- Many ergonomics standards set safety limits not only for physical loading, but also for environmental factors (Chapter 12) that may contribute further to loading of the body and mind.
- Some methods are especially targeted at manual materials handling (Chapter 10), which is a special loading situation with particular demands on workers.

Summary

- Many different ergonomics evaluation methods exist to simplify and standardize the assessment of physical loading in workplaces.
- If an evaluation is based on "rules of thumb" for what is considered good ergonomics, it is called a *heuristic evaluation*. Such an evaluation demands substantial ergonomics expertise (e.g. being a certified ergonomist or physiotherapist) on the part of the analyst to correctly and comprehensively identify risks.

- Some more formalized "checklist" and "worksheet" methods exist, many of which rely on workplace observation (either on-site or analysis of photos and video).
- Some ergonomics evaluation methods are posture-focused (e.g. OWAS, RULA, REBA, HARM), others target force exertion and biomechanical loading (NIOSH lifting equation), while others simply set acceptability limits for particular populations (Liberty Mutual materials handling tables).
- Only a few methods include time aspects like fatigue, repetitiveness and exposure time. (e.g. the Job Strain Index)
- Certain methods cover a wide range of aspects to also reflect the impact of the work environment, equipment, protective gear, etc. on the ability to perform work (e.g. KIM, RAMP, EAWS).
- Evaluation and analysis using the same method should be conducted both before and after workplace redesigns, to document and monitor progress and to enable follow-up of whether an intervention has eliminated the risk.

Notes

[1] An important condition to be aware of is that while the NIOSH load constant of 23 kg is considered safe (under ideal lifting conditions) for 99% of a male population, it is considered safe for only 75% of a female population. The root cause of this condition is that the acceptability criteria for manual lifting were originally developed to cover 90% of a working population composed equally of men and women (Waters et al., 1994 p. 759).

8.8. References

Berlin, C., Örtengren, R., Lämkull, D. & Hanson, L. (2009). Corporate-internal vs. national standard — A comparison study of two ergonomics evaluation procedures used in automotive manufacturing. *International Journal of Industrial Ergonomics, 39* (6), 940–946.

Douwes, M. & de Kraker, H. (2014). Development of a non-expert risk assessment method for hand-arm related tasks (HARM), *International Journal of Industrial Ergonomics,* 44 (2), 316–327.

Hignett, S. & McAtamney, L. (2000). Rapid Entire Body Assessment (REBA) *Applied Ergonomics* 31(2), 201–205.

IMD. (2015). Risk screening. The Ergonomic Assessment Work-Sheet (EAWS). [Online] Available at: http://mtm-international.org/risk-screening-the-ergonomic-assessment-work-sheet-eaws/ [Accessed 21 June 2016].

Lind, C. (2015). Användarmanual för Bedömningsverktyget RAMP II — Risk Assessment and Management Tool for Manual Handling Proactively, Version 2015-09-23. (in Swedish) KTH Royal Institute of Technology. [Online] Available at: https://www.ramp.proj.kth.se/ [Accessed 25 Nov 2015].

Lind, C., Rose, L., Franzon, H. & Nord-Nilsson, L. (2014). RAMP: Risk Management Assessment Tool for Manual Handling Proactively. In *Human Factors in Organizational Design And Management –*

Xinordic Ergonomics Society Annual Conference – 46 / [ed] O. Broberg, N. Fallentin, P. Hasle, P. L. Jensen, A. Kabel, M.E. Larsen, T. Weller, 2014, 107–110 s.

Louhevaara, V. & Suurnäkki, T. (1992). *OWAS: A method for the evaluation of postural load during work*. Training Publication II. Finnish Institute for Occupational Health, Helsinki.

McAtamney, L. & Corlett, E. N. (1993). RULA: A survey method for investigation of work-related upper limb disorders. *Applied Ergonomics,* 24(2): 91–99.

Moore, J. S. & Garg, A (1995). The strain index: A proposed method to analyze jobs for risk of distal upper extremity disorder. *American Industrial Hygiene Associate Journal*, 56(5): 443–458.

Niu, S. (2010). Ergonomics and occupational safety and health: An ILO perspective. *Applied Ergonomics*, 41(6): 744–753.

Rose, L. (2014). RAMP: Ett nytt riskhanteringsverktyg. Risk Assessment and Management Tool for Manual Handling Proactively (in Swedish) Project report, Dnr. 090168. KTH Royal Institute of Technology. [Online] Available at: https://www.kth.se/sth/forskning/halso-och-systemvetenskap/ergonomi/framtagna-verktyg/ramp/slutrapport-1.511645 [Accessed 25 Nov 2015].

Snook, S. H & Ciriello, V. M (1991). The design of manual handling tasks: revised tables of maximum acceptable weights and forces. *Ergonomics*, 34(9): 1197–1214.

Swedish Work Environment Authority (1998). Ergonomics for the Prevention of Musculoskeletal Disorders – Provisions of the Swedish National Board of Occupational Safety and Health on Ergonomics for the Prevention of Musculoskeletal Disorders, together with the Board's General Recommendations on the Implementation of the Provisions. Solna, Sweden: Arbetarskyddssty-relsen. Legal provision.

Swedish Work Environment Authority (2012). Belastningsergonomi — Arbetsmiljöverkets föreskrifter och allmänna råd om belastningsergonomi. ISBN 978-91-7930-565-9. [Online] Available at: https://www.av.se/globalassets/filer/publikationer/foreskrifter/belastningsergonomi-foreskrifter-afs2012-2.pdf [Accessed 25 Nov 2015].

TNO (2012). Welcome to the Hand-Arm Risk-assessment Method (HARM)! [Online] Available at: https://www.fysiekebelastingbeoordelen.tno.nl/en/page/harm [Accessed 21 June 2016].

Waters, T. R., Putz-Anderson, V., Garg, A. & Fine, L. J. (1993) Revised NIOSH equation for the design and evaluation of manual lifting tasks. *Ergonomics*, 36(7): 749–77.

CHAPTER 9

Digital Human Modeling

THIS CHAPTER PROVIDES:

- A brief overview of the function of digital human modeling (DHM) and ergonomic simulations in the design of healthy workplaces.
- A walkthrough of ergonomics analysis functions commonly found in DHM software.

How to cite this book chapter:
Berlin, C and Adams C 2017 *Production Ergonomics: Designing Work Systems to Support Optimal Human Performance*. Pp. 161–174. London: Ubiquity Press. DOI: https://doi.org/10.5334/bbe.i. License: CC-BY 4.0

WHY DO I NEED TO KNOW THIS AS AN ENGINEER?

Certain industrial sectors have strategically reached such a level of maturity regarding ergonomics that they have the means and equipment to simulate ergonomics in a computer, using a digital human model (called a manikin). Today, there are a number of commercially available 3D CAD software packages suited to ergonomic simulation, with built-in analysis functionalities that help an analyst to determine whether a new product or workplace design carries an ergonomic risk. Digital human modelling also offers the possibility to design and test new workplaces and products on a variety of virtual humans of different shapes and sizes. Knowing how to use these tools can save you time and money, allow you to compare several alternatives, and offer you better possibilities to design for a population of users, who might not be available in real life for testing.

WHICH ROLES BENEFIT FROM THIS KNOWLEDGE?

The *system performance improver* and *work environment/safety specialist* are both able to use ergonomics simulation as a way to make a business case for workplace improvements. Modelling the ergonomic risks and the impact of a design change is a fast, cheap, non-intrusive way to explain to other stakeholders what the immediate impacts of a change project would be (for example, introducing lifting equipment, changing a working height, etc.) without making changes that may disturb on-going production. The visualizations can also be used to communicate with suppliers of equipment or with operators about different risks and ways to address them.

The *purchaser* (and the *system performance improver*) can benefit from discussing a simulation of a proposed ergonomics intervention as part of the process of deciding whether the proposal is possible to fit in the available space, if it can successfully apply to all of the targeted workforce, and if the investment in e.g. new workspaces or equipment is likely to lessen the identified problematic loadings.

9.1. Ergonomic simulation

Simulations are used in many different sectors to test solutions before they are fully implemented and the discipline of ergonomics is no exception. Digital human modeling (DHM) is a term that designates a software tool that enables digital models of humans to interact with virtual workplaces or products in a digital CAD environment. The workplace can be built up in CAD and a number of tests done to determine its ergonomic suitability by importing a digital human, thus providing a visual representation of the workplace design in use. The size measurements of the digital human model are based on anthropometric databases, enabling a number of different models of different percentiles to be used in the same virtual workplace. There are a number of benefits provided by DHM for the production engineer:

- Easy to adopt a proactive design approach
- Enables numerous alternative solutions to be compared
- It is not always possible to access the real environment
- Easy to test a range of different measurements across different genders and nationalities
- Visualises the proposed work design layout and its effects on physical ergonomics, enabling meaningful discussions between designers, ergonomists and leadership
- Training aid

Given the high costs and large amount of space required to build full-scale models during the design and development stage, it can be difficult and costly to identify work tasks that may involve awkward postures or potential damaging body positions early on. Especially in the production environment where meaningful tests can only be conducted when the assembly line is shut down. Through the use of ergonomic simulations it is possible to make informed design decisions early on. Rather than only realizing there is an issue once the system has been implemented and injuries have started to occur, adopting a proactive approach and thoroughly testing and analysing design options early on through simulation can reduce injury risks and save time and money later in the implementation process.

As we have discussed throughout this book, production environments need to suit a diverse range of people with different sizes and strengths. Finding such a diverse range of people with enough time to conduct workplace testing can be difficult. However, with access to numerous databases and measurement sets, it is possible to import a range of different sized human models, of different genders and nationalities with varying percentiles using DHM. Care should be taken when using preloaded anthropometric databases, to ensure that the data is a true representation of the desired population. In addition to these testing functions, DHM can also be used as a training aid, so long before the workplace has been completed and implemented, operators can visually see what their tasks will look like.

A number of ergonomic simulation packages have been developed the last two decades, some of which are research projects and some of which are commercially available. Some noteworthy examples are:

- Jack (Siemens, 2016)
- RAMSIS (Human Solutions 2010; Bubb et al., 2006)
- SAMMIE (Marshall et al., 2010)
- DELMIA Ergonomics Specialist (Dassault Systèmes, 2016a); its well-known predecessor was DELMIA V5 Human (Dassault Systèmes, 2016b)
- Anybody (Anybody Technology, 2016)
- SANTOS (a "Virtual Soldier"; ESI Group, 2016; Abdel-Malek et al., 2007)
- IMMA (Hanson et al., 2012; 2014)

While the functionality and usability vary across the board, ultimately they all provide a method to test scenarios in a virtual environment using digital human models. Given that Siemens's software, Jack, is a well-known digital ergonomics evaluation tool that is easily researched online – and is offered with a free 30-day trial version – the majority of the terms used in this chapter will be exemplified with that software. However, many of the functions can typically be found within other programs (albeit with a different name).

9.2. Computer manikins

The human models used in ergonomic simulations are known as computer manikins and are a geometrical models of the human body that obey a set of biomechanical rules, with similar functional behaviour and capabilities as a real human. Given the complexity of the human body, with high levels of variation between individuals, it is impossible to create a truly accurate representation, so most manikins appear somewhat robotic in their movements. The computer manikins used in Jack are made up of 70 segments, 69 joints and 135 degrees of freedom. The manikins can be viewed as skeletons or as human representations as shown in Figure 9.1. Both female and male human representations exist within Jack software, Figure 9.2. Within the software it is possible to select a multitude of different sized models representing different percentile groups, Figure 9.3.

9.3. Manipulation of manikins

It is possible to move the manikins around to position them in certain ways, so that human postures and movements under certain task and environmental conditions can be simulated. In Jack this can be done in four different ways:

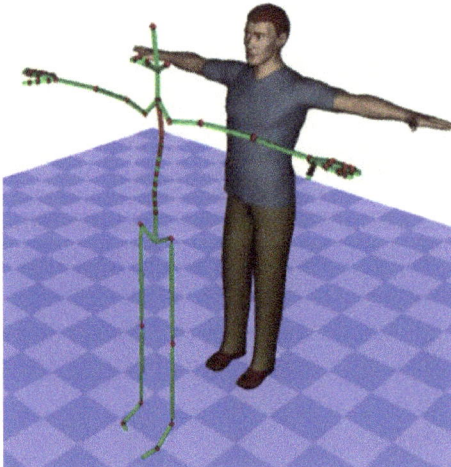

Figure 9.1: Computer manikin (skeleton and render) (Siemens, 2014).

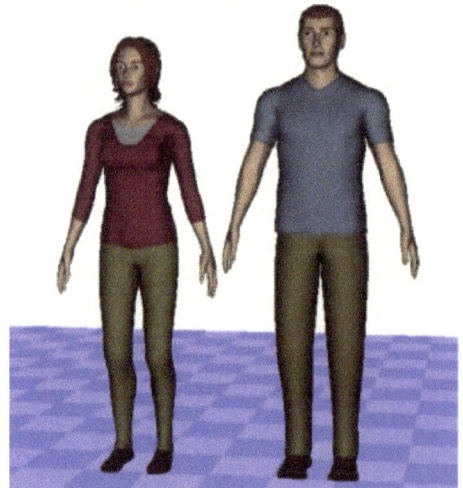

Figure 9.2: Male and female digital human models (Siemens, 2014).

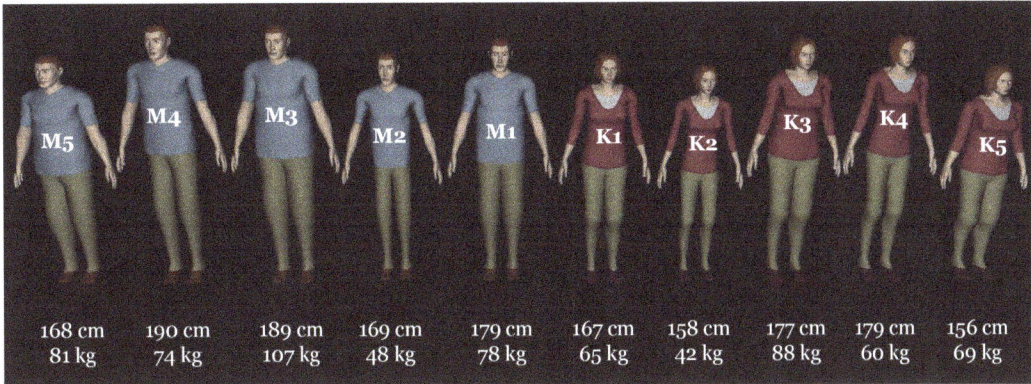

Figure 9.3: Digital human models with different percentile measurements (Brolin, 2013).

© 2016 Siemens Product Lifecycle Management Software Inc. Reprinted with permission.

Manipulating individual joints

• Using inverse kinematics
• Using pre-recorded data
• Using the Posture Wizard
• Manipulating individual joints

The simplest way to move and position manikins is to select the individual joint and drag it in the desired distance in the x, y or z axis (Figure 9.4). Ranges have been set within the software so theoretically the limbs can't be positioned outside the capabilities of the human body. However this is the most laborious method and achieving specified body postures can be very difficult.

Inverse kinematics

This is a mathematical method for controlling the movement of joints and position of the human body. While kinematics concerns calculating the position in space of the end of a linked structure, based all the different joint angles, inverse kinematics does the opposite. So the end point of the structure is known, e.g. the right thumb should be touching a button while the shoulder remains fixed, so the software will calculate what angle the joints should be at to allow this. However, given that in reality there is a range of possible solutions, in some instances the simulation might generate non-humanly possible movements based on the mathematical algorithm. Since this method can have a degree of error, combining it with manipulating individual joints can provide an accurate simulation.

Pre-recorded data

A number of common postures such as sitting and driving postures are stored within the program itself, so the user simply has to select the desired posture from a list and the manikin will automati-

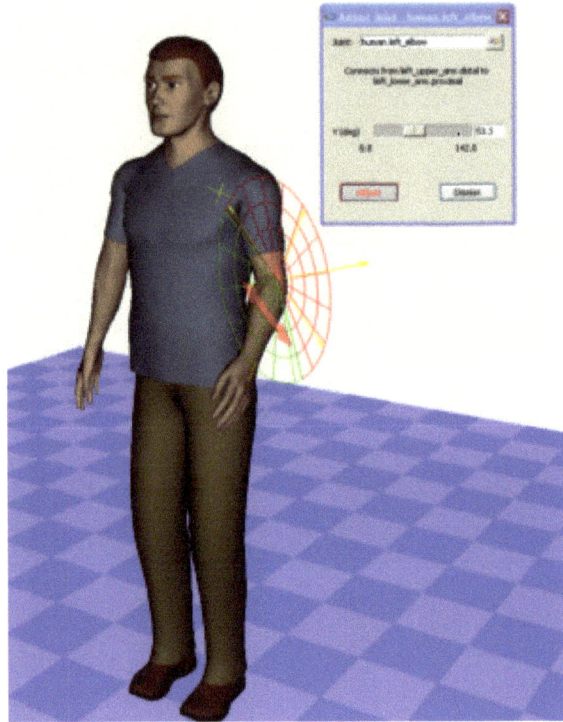

Figure 9.4: Manipulating joints (Siemens, 2014).
© 2016 Siemens Product Lifecycle Management Software Inc. Reprinted with permission.

cally adopt this position. The library of pre-recorded data is fairly extensive, with a range of different postures as shown in Figure 9.5; however, there will be some scenarios when the human model position needs to be set manually.

Posture Wizard

In this method, users set certain rules and constraints and the software creates postures accordingly. For example, in the case where an operator is leaning into a car exterior to attach the gear stick during assembly, constraints can be set regarding where both the feet and arms should be positioned so the operator can keep in balance without touching the shell. This method is considered faster and less fidgety than directly manipulating each individual joint of the manikin.

9.4. Analysis tools

In addition to manikin manipulation there are a number of other analysis tools within ergonomic simulation software that can aid in detecting risk areas early on. It is not possible or reasonable to

Figure 9.5: Pre-recorded postures in Jack (Siemens, 2014).
© 2016 Siemens Product Lifecycle Management Software Inc. Reprinted with permission.

cover all the features of ergonomic simulation tools in this chapter, so instead a brief overview of key features will be provided, to enable production engineers to grasp the bigger picture and better understand the possibilities that exist within DHM for carrying out ergonomic tests during the design phase. The following analysis tools will briefly be discussed:

- animation
- lower back analysis
- static strength prediction program
- comfort analysis
- RULA
- NIOSH
- field of View
- space and reach

(However, it is important to note that this is not an exhaustive list.)

Animation

This feature allows users to make short animations, showing the human model carrying out the work tasks, enabling potential hazards to be identified. This can either be done as a key frame animation, where the user sets up the manikin posture for each phase and the software connects the phases, creating movements and the animation. Alternatively, motion capture tools (similar to those used in animated movies) can be used. In this case a real human conducts the task while wearing a special suit and data is collected to form an algorithm enabling a digital model and animation to be made.

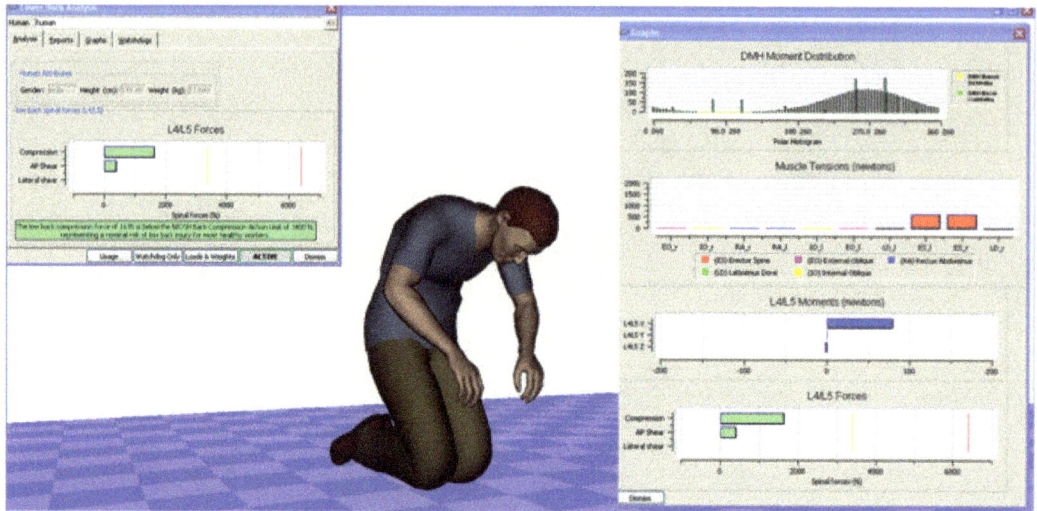

Figure 9.6: Lower back analysis tool in Jack (Siemens, 2014).

© 2016 Siemens Product Lifecycle Management Software Inc. Reprinted with permission.

Lower back analysis

This application forms a biomechanical model of the upper body of the manikin so that the static forces and torque present in the lumbar spine can be quantified. By comparing the simulated value with stored data the software can indicate whether the position is a risk or not. By having a value for the forces present in the lower back it is possible to compare alternative solutions and identify which presents the lowest level of risk. Figure 9.6 shows the output information provided by the software highlighting areas of high loading and which lumbers are experiencing the highest load.

Static Strength Prediction Program (SSPP)

This application uses biomechanical research to predict whether or not specified tasks are suitable for a certain population. This is particularly relevant for the production industry with a diverse working population of varying strength.

Comfort analysis

Another feature within DHM is comfort analysis. Given that quantifying something relatively subjective such as comfort is very dependent on posture and the environment, care should be taken when using this tool. A postural comfort metric had been established through a series of observations and empirical surveys. Responses from participants concerning comfort where collected while they were certain postures while carrying out specific tasks. These responses were then combined with joint angles measurement to create the postural comfort metric.

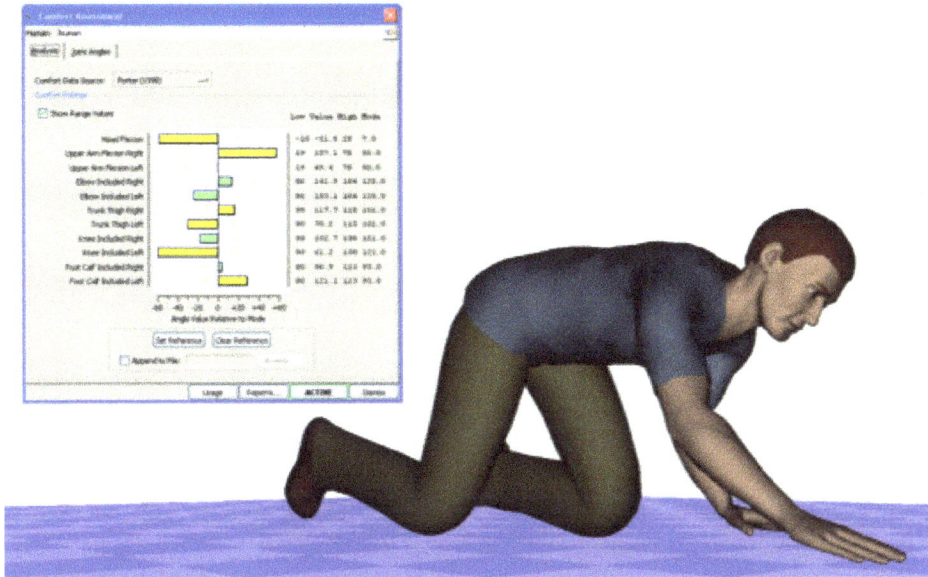

Figure 9.7: Comfort analysis tool in Jack (Siemens, 2014).

© 2016 Siemens Product Lifecycle Management Software Inc. Reprinted with permission.

By comparing these stored values with the simulated human model it is possible for the software to allocate each body section with a comfort-based score. The output of this analysis tool is shown in Figure 9.7, using colour coding to highlight instances and body parts that are regarded as uncomfortable.

RULA

DHM software also incorporates a number of ergonomic methods, some of which have been introduced in Chapter 6. For example, the built-in RULA tool can be used to identify the risk of triggering upper limb disorders in certain working postures. By inputting certain information, such as static loads along with the manikin posture the software can allocate a RULA score, indicating whether or not changes should be made to the work sequence (Figure 9.8). It is important to remember that observation-based methods, such as RULA, may be a bit "oversensitive" in DHM software, since joint angles are very exactly measured; this can sometimes result in dramatic changes in risk ratings based on very small changes in joint angles when they are near a specified angle threshold.

NIOSH

Another ergonomic evaluation method that is built into the software is the NIOSH lifting equation. By providing input data about loads, frequency and posture of lifting tasks the tool uses the NIOSH lifting equation to determine whether the lifting load is acceptable. Providing the user with an analysis summary similar to that of the above RULA analysis.

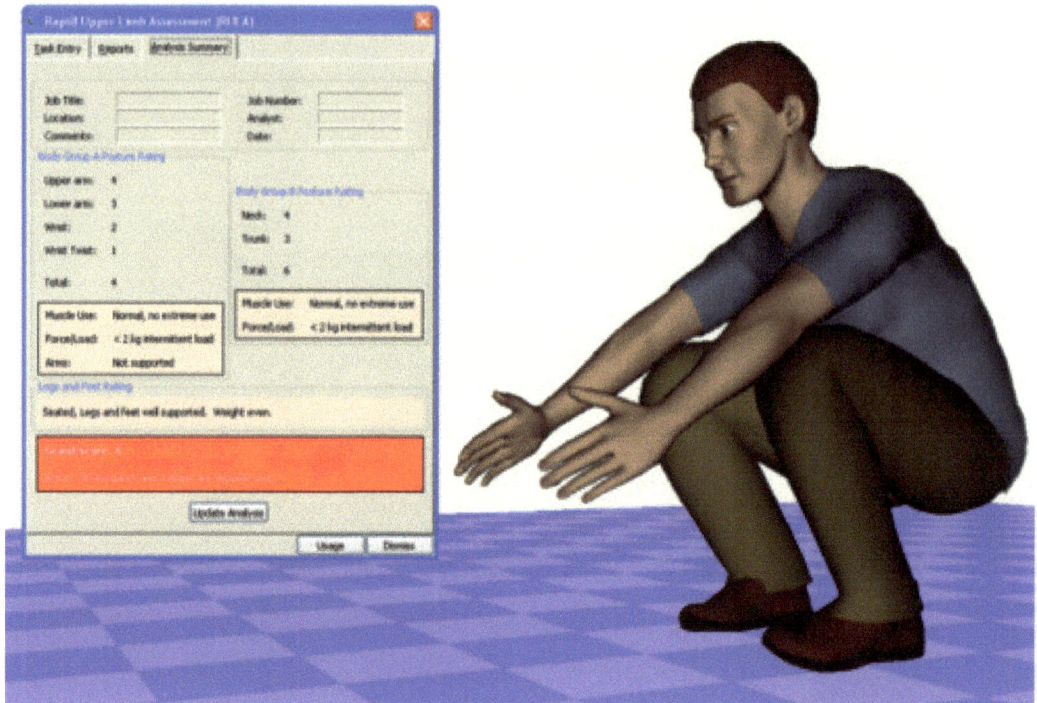

Figure 9.8: RULA tool in Jack (Siemens, 2014).

© 2016 Siemens Product Lifecycle Management Software Inc. Reprinted with permission.

Field of view

This tool provides information about the manikin's field of view, using coloured cones and boundary surfaces, a visual representation of the field of view for both eyes, the manikin's peripheral vision, their blind spot and what both the right and the left eye individually can see. Colour discrimination is also possible using this tool, so the tool gives information about areas where green, red, yellow or blue can be detected by the eye (Figure 9.9). This is particularly important when positioning warning signs. This tool makes it easy to check if screens or other necessary objects are directly in the manikin's field of vision, or if an alternate posture has to be adopted to see something.

Space and reach

The reach tool enables the areas of maximum and comfortable reach for each manikin to be easily visualized. This tool is particularly useful when ensuring the design is suitable for a diverse population and checking that all workers can components such as controls, pedals and levers. It can also be used to conduct accommodation studies to check that all the joints are provided with sufficient clearance zones.

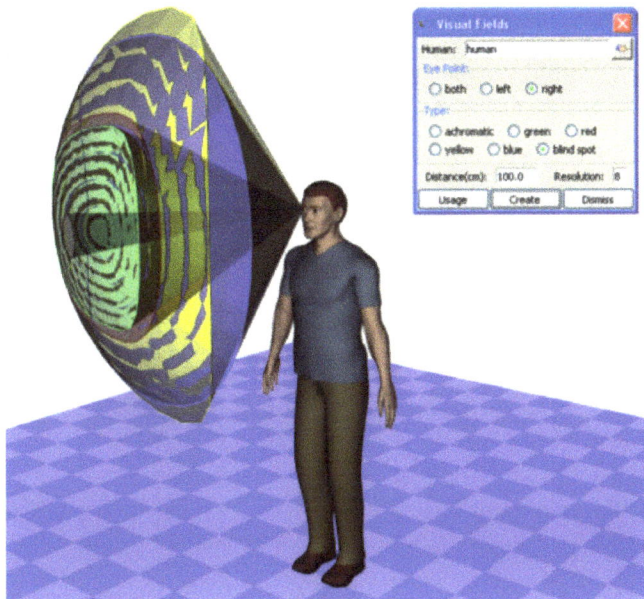

Figure 9.9: Field of view tool in Jack (Siemens, 2014).

Study questions

Warm-up:

Q9.1)   Name three reasons to use ergonomic simulation with DHMs before implementing a change to a workplace.

Q9.2)   Name some common functionalities in DHM that can be useful as decision support in workplace design.

Q9.3)   How are workforce populations represented in DHM?

Look around you:

Q9.4)   The next time you encounter a 3D digital representation of a work environment or factory, determine if there is a human representation in it to indicate scale, and possibly usage scenarios (e.g. a story of what the operator needs to do and how that action proceeds). Does the human representation succeed in illustrating potential demands or problems in the workplace? Is it possible to assess the safety and suitability of the workplace to different worker sizes?

Connect this knowledge to an improvement project

- When considering alternatives for a workplace design or re-design, DHM can be used to evaluate a 3D CAD model of the current or imagined environment to identify if particular tasks are supported or hindered by the design.
- The DHM allows the design to be tested on various manikin sizes (i.e. a *manikin family*) in order to determine whether the design is acceptable for the range of imaginable workers.
- The analysis methods and tools included in DHMs provide a quantification of the risk levels when different tasks and postures are carried out in the work environment.

Connection to other topics in this book:

- Basic biomechanics (Chapter 3) make up the basis for many computer manikins and their associated analysis capabilities. Knowing basic biomechanical principles can increase the understanding for (and critical evaluation of) DHM evaluation results.
- Knowing a bit about anthropometry (Chapter 4) and representative populations can aid the workplace designer in using DHM wisely as a design and testing tool.
- DHMs often include some ergonomics evaluation (Chapter 8) calculations as part of their analysis capabilities. Knowing how the most common ones are calculated based on observation data also increases the analyst's capacity for critical evaluation of the DHM results.
- In some companies, it may be well accepted to use DHM as a tool for involving stakeholder/worker input (Chapter 6) by providing a visualization of the workplace, work tasks and improvement proposals.

Summary

- DHM tools enable a proactive approach to ergonomics by enabling virtual testing of designs early on.
- Using human manikins with various sizes and characteristics saves companies both money and time when considering various design alternatives.
- A range of different analysis methods can be conducted within the software to test the design's suitability, including injury risk, user comfort, reachability and line of sight.
- It is important to remember that observation-based methods, such as RULA, may be a bit "oversensitive" in DHM software due to the exact interpretation of the joint angles, meaning that some postures may be rated too severely, as the rating scales are discrete rather than incremental.

9.5. References

Abdel-Malek, K., Yang, J., Kim, J. H., Marler, T., Beck, S., Swan, C., Frey-Law, L., Mathai, A., Murphy, C. & Rahmatallah, S. (2007) Development of the virtual-human SantosTM. In *International Conference on Digital Human Modeling* (pp. 490–499). Springer Berlin Heidelberg.
Anybody Technology. (2016). Products & Services. [Online]. Available from: http://www.anybodytech.com/ [Accessed 27 July 2016].
Brolin, E. (2013) Anthropometry. [Lecture] Chalmers University of Technology, 18th February 2013
Bubb, H., Engstler, F., Fritzsche, F., Mergl, C., Sabbah, O., Schaefer, P. & Zacher, I. (2006). The development of RAMSIS in past and future as an example for the cooperation between industry and university. *International Journal of Human Factors Modelling and Simulation,* 1: 140–157.
Dassault Systèmes. (2016a). Ergonomics Specialist. [Online]. Available from: http://www.3ds.com/products-services/delmia/capabilities/ergonomics/ergonomics-specialist-1/ [Accessed 27 July 2016].
Dassault Systèmes. (2016b). Human Workplace Design and Simulation. [Online]. Available from: http://www.3ds.com/fileadmin/PRODUCTS/DELMIA/PDF/Brochures/delmia-human-workplace-design-simulation [Accessed 27 July 2016].
ESI Group. (2016). SANTOS [Online]. Available from: http://soldier.esi-group.com/Subpages/santos.html [Accessed 27 July 2016].
Hanson, L., Högberg, D., & Söderholm, M. (2012). Digital test assembly of truck parts with the IMMA-tool-an illustrative case. *Work,* 41(Supplement 1): 2248–2252.
Hanson, L., Högberg, D., Carlson, J. S., Bohlin, R., Brolin, E., Delfs, N., Mårdberg, P., Gustafsson, S., Keyvani, A. & Rhen, I.-M. (2014). 'IMMA-intelligently moving manikins in automotive applications' in *Proceeding of ISHS 2014, Third International Summit on Human Simulation.*
Human Solutions. (2010). RAMSIS (version 3.81.31). [Software] Kaiserslautern, Germany: Human Solutions GmbH.
Marshall, R., Case, K., Porter, M., Summerskill, S., Gyi, D., Davis, P. & Sims, R. (2010). HADRIAN: a virtual approach to design for all. *Journal of Engineering Design*, 21: 253–273.
Siemens. (2014). Jack (Version 7-1) [Software] Siemens Product Lifecycle Management Software Inc.
Siemens. (2016). Jack and Process Simulate Human. [Online]. Available from: https://www.plm.automation.siemens.com/en_us/products/tecnomatix/manufacturing-simulation/human-ergonomics/jack.shtml

Bibliography

Siemens. (2014). Jack (Version 7-1) User Manual [Software] Siemens Product Lifecycle Management Software Inc.

CHAPTER 10

Manual Materials Handling

THIS CHAPTER PROVIDES:

- Descriptions of manual materials handling principles.
- Some examples of logistics solutions.

How to cite this book chapter:

Berlin, C and Adams C 2017 *Production Ergonomics: Designing Work Systems to Support Optimal Human Performance.* Pp. 175–188. London: Ubiquity Press. DOI: https://doi.org/10.5334/bbe.j. License: CC-BY 4.0

WHY DO I NEED TO KNOW THIS AS AN ENGINEER?

In most cases of manual assembly, the materials that become the actual product need to be transported to and from the workplace we are designing. While logistics and manual materials handling is a large, well-researched subject that deserves a book of its own, it is helpful for a production engineer to be aware of some basic principles of manual materials handling. This knowledge includes principles of packing, delivery, presentation in a particular order, and some modern industrial "tips and tricks" that are related to the knowledge you gained about cognitive aspects in Chapter 5.

Knowing about material flows is a good way to get ideas for presenting components to a worker – this may not only relieve them of unnecessary physical loading, but may also serve to improve understanding of how to assemble the product. At the same time, environmental or economical demands may sometimes dictate that a particular material flow should be used, and in such cases it is helpful for a production engineer to understand whether a) human concerns are the priority, and therefore the method of materials delivery should be changed, or b) if the other demands on the total production system may be of a higher priority, meaning that the workplace itself should be designed to work around the chosen materials handling principle.

WHICH ROLES BENEFIT FROM THIS KNOWLEDGE?

Materials Handling may traditionally be seen as the domain of logisticians and production planners – naturally, this means that it overlaps with many concerns of workplace design, in particular for assembly line settings. For this reason, many different roles may be interested in the economical aspect connected to the use of floor space and set-up time. The *manager/ leader* and the *purchaser* may have a basic understanding of certain materials principle and may regard this as just another investment but should know the relative pros and cons of choosing one over the other, as there is no single, superior method, and all of them bring benefits that come at different costs, from different system perspectives. The *system performance improver* and *work environment/safety specialist* need to know the impacts on performance, efficiency and worker safety for each materials feeding principle, so as to know how to reason about impacts and turn a proposal into a business case. The *sustainability agent* is likely to be concerned with issues of materials handling from an environmental perspective (material use and scrap avoidance), so having knowledge of the different MMH principles' social and economical impacts helps to integrate the knowledge into a sound sustainability assessment of how to supply material.

10.1. Function of manual materials handling

Lean production is all about maximizing value through the optimization of flow and elimination of waste. One way to reduce waste is to decrease the amount of handling associated with materials and parts as they go through the production chain (Jonsson et al., 2004) – i.e. *materials handling*. Reducing the amount of handling not only improves efficiency and delivers cost savings, but also provides a more operator-friendly work environment, with fewer risks that could lead to MSDs. While there are many forms of handling equipment in existence, like conveyor belts, picking robots, trucks and trolleys, a large number of manual activities are still being performed in the production environment. When material is handled without the aid of automated devices, it is classed as manual materials handling (MMH).

Materials handling is defined as:

> "The movement of raw materials, semi-finished goods, and finished articles through various stages of production and warehousing" (Compton, 1998).

While the logistics of transporting parts to and from the production site come under the broad umbrella of materials handling activities and need to be addressed by the planner, this book will not cover the logistics associated with supply and delivery. We will instead focus on materials handling from an internal factory perspective – that is, the nature of material flow within the factory, surrounding the assembly line and the tasks closely related to assembly activities. Material flow, container positioning, the supply of components to the right workstations when needed and the design of storage containers are all key characteristics of materials handling.

10.2. Issues and risks arising from poorly designed MMH systems

The design of workstations and assembly lines with regards to MMH has a direct impact on both the product's time-to-market and the health of the operator. The overall aim of the workstation is to ensure a high rate of productivity is achieved and customer demands are fulfilled without creating any unnecessary strain on the operator. Part of the responsibility for creating such workplaces falls on production engineers.

In this book we discuss the impact of loading on the body (Chapters 2 and 3) and the negative impact some work environments can have on the human body (Chapter 12), leading to the onset of MSDs. The materials handling sector is found to be a large contributor and responsible for causing many lower back related injuries, and hence absenteeism (see Chapter 11). In fact, the problems associated with materials handling are what led to the creation of the NIOSH analysis method that was introduced earlier in Chapter 8. Bending over to pick material out of pallets or picking up pallets or large heavy components is a common injury trigger (Neumann et al., 2006).

Internal factory MMH activities that are associated with assembly exist in the following forms:

- administration activities (e.g. auditing and reporting)
- taking necessary components out of storage containers
- putting components into packages

- downsizing or re-packaging components
- kitting and sequencing of components
- delivering the components to specific assembly workstations on the production line
- assembling components to form sub-assemblies
- transporting sub-assemblies down the production line to the next workstation
- combining sub-assemblies to form larger sub-assemblies or the final product
- packaging of sub-assembly or product
- disposal and handling of packaging material

Each of these tasks can involve walking large distances, lifting, pushing, pulling or carrying heavy loads, and highly repetitive movements. Given the diverse nature of all these tasks, a number of different considerations should be taken into consideration in parallel when designing the workplace. Otherwise, there is a risk that optimization efforts will be targeted only at a micro level, but the overall system performance will end up being sub-optimal (Hanson, 2012). There is also a danger that system efficiency is considered more important[1] than the operator's safety and well-being, leaving them with physically demanding tasks in obscure body positions. In line with the lean approach to production, walking to collect necessary components and handling them is considered non-value adding work, and therefore efficient materials handling aims to combat "unnecessary movement". So to reduce such "waste", production facilities strive towards workstation and assembly line layouts where the need for the operator to move away from their assembly station is eliminated, or at least minimized.

10.3. Different types of MMH

Getting the right part in the right place, with the right orientation, at the right time is crucial to efficient production both in terms of time and cost. The characteristics of the products and individual components being manufactured can have a profound impact on the nature of how they are handled. Size, weight, shape, product variants, surface finish and demands of components are all product attributes that have an impact on materials handling (Hanson, 2012). Products with complex shapes involving hooks or springs are more likely to get tangled with other components, adding additional non-value adding actions to the work of the operator. Products on the same assembly line may also come in different variants. So while the core components remain the same, some subtle variations may mean that some parts will differ – and consequently, so will the assembly sequence. In some instances, this can involve workers sorting through numerous versions of the same basic component in search of the suitable one for the product variant in question. Ensuring that workers can pick the right parts for different product variants as quickly as possible – with the least number of touches and without overloading their body – is a responsibility for the workplace designer. Large or heavy components can be difficult to handle and move, which in some cases presents a higher level of injury risk for the worker. In comparison, fragile components with sensitive surfaces that can become easily damaged will require additional care (and possibly additional physical strain) during assembly activities.

One way to speed up the MMH process and ease the work of the operator is to take away the need for them to search and choose; if all the necessary components for that product variant are laid out before them in the right quantities, the need to choose the right component from the parts shelf is eliminated. Another key constraint in factories is that space to store material containers is at a premium, particularly near the assembly line. This has often caused companies to adapt their materials handling processes to target economic efficiency.

The following four materials feeding methods are the main conventional ways to supply material:

• Line stocking
• Batching
• Sequencing
• Kitting

Deciding which materials feeding system approach to adopt is a key concern for all production facilities, which directly impacts the nature of the assembly tasks conducted and the assembly lines' overall performance. Each method will be discussed in more detail below; however, the main emphasis will be on kitting.

10.4. Line stocking

Line stocking is one of the more traditional materials feeding methods. It is also referred to as bulk feeding, continuous replenishment or point-of-use storage systems (Limère et al., 2011). Generally these systems have some degree of automation, are fairly inflexible and are specialized to a certain product, only allowing for a low degree of product variation. In line stocking, full containers of each component type are delivered in bulk to the assembly line in the same containers that they were shipped in from the suppliers (Hanson & Brolin, 2013). Each component type is stored in a different container, and all the containers are stored next to the assembly line.

Benefits of line stocking include:

• No need to pre-process or re-arrange parts.
• Stock is continually available at the assembly line (depending on the replenishment method).
• If a part is defective, it is very easy to select a new one from the container.

Disadvantages of line stocking include:

• Capital is tied up in stock.
• The shop floor becomes crowed with containers and pallets.
• Lack of space both at the production area and workstation, especially if many product variants are produced.
• A lot of time is spent walking, removing packaging, searching for and collecting the correct parts in the right quantities.

10.5. Batching

Batching is commonly used when products are made in response to a specific customer order. Typically this method is used when there is a mid to low volume of products required, with a limited variation in product type. The material is provided for a specified number of objects that are to be assembled.

10.6. Sequencing

In the sequencing method, the number and type of parts needed for a specified number of objects to be assembled are displayed at the assembly station. For situations when only a small number of components are assembled per station, this method is recommended rather than kitting, as there is little value in taking time to make kits for each station when only a few components are necessary.

10.7. Kitting

Kitting is the term applied to the practice of collecting a predetermined amount of material in the form of components and/or sub-assemblies in containers and delivering them to the assembly line, to support one or more assembly operations for a given product or shop order (Bozer and McGinnis, 1992). The containers or bins storing the necessary material are known as kits. Kitting essentially splits the assembly process into two distinctive phases; the collection of the required material, followed by the part assembly. Kitting can be done in two forms: a travelling kit or a stationary kit (Bozer and McGinnis, 1992). A travelling kit is one that moves down the assembly line at the same time as the object being assembled, so each kit contains material for several different workstations. Stationary kits, on the other hand, only contain the necessary parts for one specific workstation.

Due to the nature of the product and factory layout, kitting is a necessity in some industries – however, it is important to note that kitting is not necessarily the superior choice and it is rarely implemented from the outset. In an ideal world, a product would have a relatively low number of components (based on Design For Assembly, DFA principles) and be designed in such a way that makes it simple to assemble; however, this is rarely the case in reality. So in some circumstances, kitting is necessary and can be used to solve the following issues, which will be elaborated in the following sections:

- Lack of space
- Quality
- Flexibility
- Materials handling
- Learning

Lack of space

A common problem with line stocking is the amount of space required to store the large material containers for each component type next to the assembly line (Finnsgård et al., 2011). Kitting removes the need to have material pallets and containers located right beside the assembly line by only providing operators with the amount of material required to assemble one product/sub-assembly at a time, thus freeing up space on the shop floor. In reality, this means space is taken up elsewhere in the factory upstream of the assembly line, as the material still needs to be stored and a space needs to be allocated for the making of kits in the first instance. So in a holistic perspective, kitting can actually take up more space, since the original material containers still exist along with the new kitting containers — however, the value of space is relative to its proximity to the line itself, so it can still make good economical sense to devote space to kitting elsewhere. This liberation of more space by the assembly line is often used as the number one business case when kitting is implemented in Swedish factories (Corakci, 2008). Internationally, kitting is most often used to enable mixed model assembly lines in automotive assembly. See also Flexibility.

Quality

As previously mentioned in Chapter 5, kitting can also contribute to enhanced quality. Since the operator does not need to focus on what parts to assemble, they can instead focus all their energy and efforts on the assembly tasks (Bäckstrand, 2009; Medbo, 2003). Kits also provide a memory aid to operators by clearly showing if any components have been missed or forgotten during assembly, so issues can be corrected as soon as they occur. To ensure high quality and minimize confusion, kitting trays should have both component-shaped holes indicating where everything belongs and part identification numbers that correspond to the large bulk storage containers. Having inserts in kitting containers that prevent component movement can also protect sensitive surfaces from getting damaged (Corakci, 2008).

Flexibility

Kitting is considered to be a more flexible materials supply medium than line stocking (Sellers & Nof, 1986; Bozer & McGinnis, 1992). Providing material in kits rather than large storage containers facilitates the making of product variants at any workstation, since the issue of space limitation is removed. Kitting also encourages operators to be more flexible as it better equips them to assemble different product types when all components are presented in a logical manner.

Materials handling

In the case of kitting the total materials handling time can be seen as the sum of materials handling by the assembler, materials handling during kit preparation, and internal transportation to get the parts to the assembly line. The use of kitting is believed to save time as the need for the assembler to walk around, search through containers and collect material for each individual product is eliminated. Theoretically, these accumulated time savings can be transferred to justify the employment of a kitter (although other staffing management issues may limit this possibility).

Kitting also means there is less variation in the time to complete tasks, as handling and walking can significantly vary between workers dependent on individual strength, body size, etc. Most kit preparation stations involve some sort of picking support system, e.g. pick by light or voice, to ease the cognitive load on the operator and maximize productivity. Kitting is somewhat limited to small or medium sized components, since some larger components can't fit into a generic kitting container. Having said that, in some cases rack systems are used rather than containers for large components. In such instances a hybrid of line stocking and kitting is used. Typically, fasteners aren't included in kits as they are frequently used on many product variants, can easily be dropped and lost (which would be a problem if the assembler is supplied with an exact number of them), and don't require a large volume of space to store in bulk (Hanson & Brolin, 2013). Alternatively, fasteners can be stored in small containers beside the power tools so operators can just grab a handful of them when collecting the tools.

Learning

As previously discussed in Chapter 5, kitting aids the operator by positioning the material in the container in such a way that it acts as a work instruction, showing the assembly order. Implementing standardized work associated with kitting also provides benefits from the assembler's perspective. The holistic learning strategy provided by kitting leads to shorter learning times overall during the introduction of new product variants, or for new employees (Medbo, 1999).

Additional benefits that kitting can bring:

- pace keeper (takt time)
- facilitated materials control (components are at hand)
- better visibility of the shop floor and assembly line flow

Kitting summary

While kitting provides a number of benefits when successfully implemented, it is also important to be aware of its weaknesses and the potential issues it can cause. The success of both kitting and continuous flow is very much dependent on the setting and exact way in which these processes are applied.

Often, containers are designed in such a way that the parts can only be positioned one way, making it easier for both the kitter and the assembler. However, in situations where there is a high degree of product variance, there will be a need for a number of different containers. The biggest problem is when the product design is changed, requiring a different sequence or components, in such a way that the containers become obsolete and have to be redesigned. In some cases kitting can lead to a higher number of man-hours, as the material has to be kitted, taken to the assembly line and then assembled (Sellers & Nof, 1986; Bozer & McGinnis, 1992). So it is only worthwhile to implement kitting if it is expected to provide numerous benefits in other areas that balance this out – such as higher quality, increased flexibility (product mix, new products, volume flexibility, changed takt time), higher production capacity, reduced time variation between assemblers, and the provision of more space on the shop floor.

If a component is missing from a kit, a replacement is sometimes "stolen" from another kit, which can lead to shortage issues in kits further down the line and increase the amount of handling. In situations where parts can't be taken from another kit, it may be necessary for the operator to walk all the way to the kitting area and retrieve the necessary component. This slows down the whole assembly process and in some cases can force stoppages. In contrast, such an issue wouldn't occur if continuous flow were used, as the distance from the storage container to the assembly line is significantly shorter (Hanson & Brolin, 2013). Kit preparation also means that parts are handled an additional time, which in the case of fragile or sensitive parts increases the risk that they will become scratched, warped or damaged. Like the implementation of any new change in a production environment, kitting has costs associated with its introduction, as the setup of any new process will require layout modifications, new equipment and staff training. In instances where kitting is determined as the best option, managers should be wary of "over- kitting" and kitting unnecessary parts. Kitting is often used in parallel with line stocking, where high-variation products are kitted and all the other parts are line-stocked.

10.8. Workstation design principles

Given that materials handling affects both the physical workload of operators and the performance of the system, it is worth outlining some key guidelines and principles concerned with the design and layout of workstations.

The following principles will be discussed in more detail below:

• Working height considerations
• Storage container considerations

10.9. Working height considerations

The back and the wrists are often at risk in materials handling tasks. With well thought-out designs and layouts such risks can be minimized, enhancing the work environment for the operator. Removing material from containers often involves bending or stretching and is identified as a high-risk activity, especially given the high degree of repetition associated with the collection of material and assembly activities.

The height and angle of the shelves storing the containers, and their location relative to the assembler, are a key design consideration. The frequency of tasks involving back bending and arm raising should be identified and designs put in place to reduce or eliminate these high injury-risk occurrences.

Kitting can somewhat lessen bending motions, as components are stored in the kitting container which is then positioned at a suitable height relative to the worker and the workstation. Some workstations are even large enough that kitting containers can be stored on the same surface where assembly operations are conducted – however, this is not always the case, and is very dependent on product and part size. In situations where there is limited space preventing the operator from accessing all the necessary material from their normal working position, a common guideline is that the most frequently used material should occupy the best position. Generally, taking a step to collect material is considered better for the body than staying in the same place and twisting or bending to access the components.

While the use of such containers and layouts have potential to eliminate undesirable body postures and high loads, the time saving could trigger a phenomenon dubbed the "ergonomics pitfall" (Westgaard & Winkel, 1997), which means that time gains are used to spend even more time working (e.g. on assembly), which in turn could lead to an increased risk of repetitive strain injury. If this intensification of work occurs unintentionally, it may introduce its own new risk factors and problems, and the ergonomic benefit that was meant to increase safety and reduce risk is lost.

Given that operators have different body characteristics and work at numerous different stations, it can be beneficial to have containers stored on height-adjustable shelves, which the operator can set to an appropriate height at the start of their shift. However, this obviously increases equipment set-up cost and some time will be taken every time the height is adjusted, so a review would be necessary to check whether the gains from such a system (e.g. decreased operator strain and time savings) justify the setup.

Some companies have been known to develop their own storage design guidelines that strive to lessen the physical strain of materials handling — one example is Volvo's VASA model (Backman, 2008; Finnsgård et al., 2011), which is based on acceptability criteria values from Volvo's own corporate standard (Volvo, 2014); see Figure 10.1. By extension, its "translation" into material façade risk zones (by Finnsgård et al., 2011) is shown in Figure 10.2.

Components stored at heights that cause a high load on the body are given a red classification; typically, these are very high up or very low down. Whereas components that are easy to collect without causing any discomfort are given a green classification. Based on this model, container layouts are normally set so that the most frequently used are easiest to obtain.

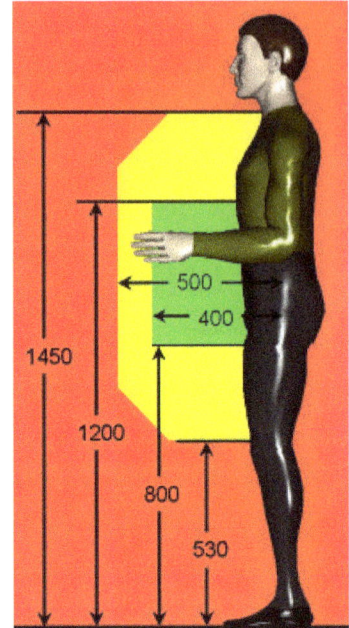

Figure 10.1: Criteria levels from Volvo's internal ergonomics standard (Volvo, 2014), commonly applied at numerous Volvo production facilities. Exposed material is classed into one of three categories (red, yellow or green) based on the lifting frequency impact on an "average" 172 cm tall person.

Image reproduced with permission from B. Johansson/Volvo. All rights reserved.

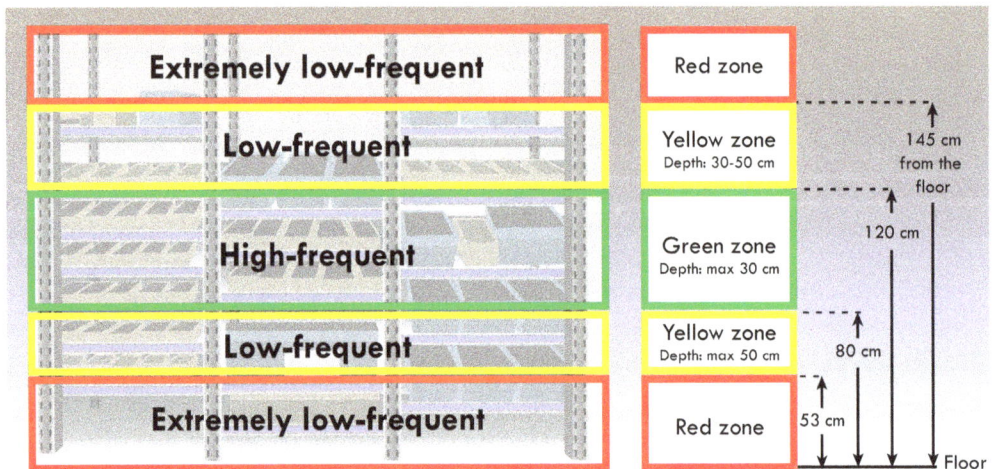

Figure 10.2: The VASA model translated into red, yellow and green zones for a material façade, indicating how frequently it is acceptable to handle materials at each height.

Image by C. Berlin, based on Finnsgård et al. (2011).

Figure 10.3: Gravity flow rack and kitting containers.
Photograph by C. Adams.

10.10. Storage container considerations

The sizing and positioning of storage containers also contributes both to the efficiency of materials flow and the injury risks for assemblers. Material containers are stored on a shelving rack, also known as a *façade*.

A study (Neumann & Medbo, 2010) was conducted comparing the use of large containers (Euro pallet sized, 800 mm x 1200 mm) to narrower boxes. Through a biomechanical analysis studying the loads the human body is subjected to during work, it was identified that narrow boxes reduce both peak spinal loads and shoulder loads. The study also showed that the use of narrow boxes enabled shorter material supply racks to be used, providing more space next to the assembly line and decreasing the overall assembly time by reducing the amount of walking needed to collect parts. Smaller boxes also enable parts to be stored closer to each other, leading to less time being spent walking and collecting material – this is a key benefit for scenarios involving many product variants, with numerous material containers. The use of narrow boxes stored in racks on wheels[2] provides a higher degree of flexibility, as the whole rack can simply be rolled to a new location if changes are made to the layout.

Another study (Finnsgård, 2013) has confirmed the benefits of smaller storage containers, reporting a 23% reduction in materials handling time after a redesign was made utilizing smaller containers. The same study also identified a 67% reduction in space needed to expose the components, resulting in substantially more space next to the assembly line. Using open-fronted containers or angling them towards the operators also provides benefits, such as making it easier to see exactly what is in the container (reducing neck strain) as well as improving accessibility, reducing unnecessary wrist extension and flexion. Gravity flow racks are often used to hold storage containers and aid accessibility (Figure 10.3). These inclined shelves mean that when the front container is taken out, the next container will move forward to the front of the rack. They also enable kitters and assemblers to work simultaneously without getting in each other's way, as the kitter can stock the containers from the back while the assembler collects material from the front.

Study questions

Warm-up:

Q10.1) Name at least three risks for quality and safety that can stem from poor MMH ergonomics.

Q10.2) What are the pros and cons of kitting?

Q10.3) What are the pros and cons of line stocking?

Q10.4) What are the pros and cons of using small containers?

Look around you:

Q10.5) What are some good solutions for MMH that you have seen in industrial or retail settings? Think of examples where stocks need replenishment fairly frequently.

Q10.6) Observe the physical arrangements of the goods in a warehouse store (i.e. a self-service shop where customers fetch goods spread out over a large area), for example a home improvement store. Would you say that the store has provided kitting or sequencing solutions to enable customers to easily acquire all the materials and tools they need to assemble goods at another location, or do customers need to visit several different areas of the store to fetch all they need?

Connect this knowledge to an improvement project

- MMH may sometimes be regarded as a separate area of responsibility from workplace design, but taking it into consideration helps to ensure that materials presentation is appropriate for assemblers, and that the needs and limitations of materials handlers are included in design proposals.
- MMH principles like kitting, batching, etc. can be used as "solution kits" to address physical, cognitive and psychosocial aspects of ergonomics, since component visibility, time-keeping, quality control and teamwork may all be positively impacted by appropriately chosen MMH solutions.

Connection to other topics in this book:

- Some ergonomics evaluation methods (Chapter 8) are specifically targeted towards MMH (see section 8.2.3), e.g. KIM and the Liberty Mutual manual materials handling tables.
- Many MMH principles are based on avoidance of excessive physical loading (Chapter 3) and sound cognitive ergonomics principles (Chapter 5).
- The economics of a workplace (Chapter 11) may be positively impacted by well-designed MMH – not only due to fewer injuries and increased speed and efficiency, but also because some MMH solutions can free up valuable floor space.

Summary

- Materials handling systems have a significant impact on the performance of assembly systems.
- Use a holistic view, weighing up all the trade-offs when selecting which materials handling method to use.
- Four main methods in operation: line stocking, sequencing, batching and kitting.
- Kitting provides benefits in the areas of quality, lack of space, learning, materials handling and flexibility.
- Workstations should be designed taking into consideration walking distances, positioning of components and storage containers relative to the assembler and the frequency of component selection.

Notes

[1] Of course, this misconception can be best avoided by adopting the design philosophy that well-being and system efficiency are mutually compatible, and that solutions can be found to benefit both.
[2] For safety reasons, these should of course be wheels that lock.

10.11. References

Bäckstrand, G. (2009). Information flow and product quality in human based assembly. PhD thesis, Loughborough University, UK.

Backman, K. (2008). VASA Ergonomic requirements – Volvo Corporate Standard STD 8003, 2.

Bozer, Y. A. & McGinnis, L. F. (1992). Kitting versus line stocking: A conceptual framework and a descriptive model. *International Journal of Production Economics,* 28: 1–19.

Compton's Interactive Encyclopaedia. Cambridge, Mass: Softkey Multimedia, 1998.

Corakci, M. A. (2008). An evaluation of kitting systems in lean production. Master's thesis. University of Borås, Sweden.

Finnsgård, C. (2013). Materials exposure: The interface between materials supply and assembly. PhD thesis, Chalmers University of Technology, Sweden.

Finnsgård, C., Wänström, C., Medbo, L. & Neumann, P. (2011). Impact of materials exposure on assembly workstation performance. *International Journal of Production Research,* 49(24): 7253–7274.

Hanson R. (2012). In-plant materials supply: Supporting the choice between kitting and continuous supply. PhD thesis, Chalmers University of Technology, Sweden.

Hanson, R. & Brolin, A. (2013). A comparison of kitting and continuous supply in in-plant materials supply. *International Journal of Production Research,* 51(4): 979–992

Jonsson, D., Engström, T., & Medbo, L. (2004). Some considerations relating to the reintroduction of assembly lines in the Swedish automotive industry. *International Journal of Operation & Production Management,* 24(8): 754–772.

Limère, V., Van Landeghem, H., & Aghezzaf, E.-H. (2011). Kitting versus line stocking in the automotive assembly industry: the influence of part characteristics, In: Belgian Operations Research Society, 25th Annual conference. 147–148.

Medbo, L., 1999. Materials Supply and Product Descriptions for Assembly Systems. PhD thesis, Chalmers University of Technology, Sweden.

Medbo, L., (2003). Assembly work execution and materials kit functionality in parallel flow assembly systems. *International Journal of Industrial Ergonomics,* 31(4): 263–281.

Neumann, W. P., Winkel, J., Medbo, L., Mathiassen, S. E. & Magneberg, R. (2006). Production system design elements influencing productivity and ergonomics – A case study of parallel and serial flow strategies. *International Journal of Operations & Production Management,* 26(8): 904–923.

Neumann, W. P. & Medbo, L. (2010) Ergonomic and technical aspects in the redesign of material supply systems: Big boxes vs. narrow bins. *International Journal of Industrial Ergonomics,* 40(5): 541–548.

Shtub, A. & Dar-El, E. M. (1989). A methodology for the selection of assembly systems. *International Journal of Production Research,* 27(1): 175–186.

Sellers, C. J. & Nof, S. Y. (2007). Part Kitting in Robotic Facilities In: *Robotics and Material Flow.* Amsterdam: Elsevier Science Publishers B. V. pp. 163–174.

Volvo (2014). Ergonomic requirements – Application. Volvo Corporate Standard STD 8003, 29. [Online]. Available from: http://www.volvogroup.com/en-en/suppliers/useful-links-and-documents/corporate-standards/search-for-standards.html [Accessed 12 Dec 2016].

Westgaard, R. H. & Winkel J. (1997). Ergonomic intervention research for improved musculoskeletal health: a critical review. *International Journal of Industrial Ergonomics,* 20(6): 463–500.

CHAPTER 11

The Economics of Ergonomics

Image reproduced with permission from kurhan/Shutterstock.com. All rights reserved.

THIS CHAPTER PROVIDES:

- Design concerns when addressing the well-being, performance and retention of future workforce.
- Case studies showing how ergonomics and economics are interlinked.
- Descriptions of stakeholder relations and persuasive behaviours.
- Some cost-related calculation procedures to evaluate an ergonomics investment.

How to cite this book chapter:
Berlin, C and Adams C 2017 *Production Ergonomics: Designing Work Systems to Support Optimal Human Performance.* Pp. 189–212. London: Ubiquity Press. DOI: https://doi.org/10.5334/bbe.k. License: CC-BY 4.0

WHY DO I NEED TO KNOW THIS AS AN ENGINEER?

Quite frequently, the people in a company who know the most about human factors and ergonomics are not the people who "own" the design problem or the design of assembly solutions; that is, they do not have the right themselves to change the workplace. This is especially true for ergonomists – for many of them, it becomes absolutely essential to develop interpersonal skills and a "language of economics" so that they can use cost-related persuasive arguments when communicating with people who have the mandate to put money and resources towards making a change (for example engineers, production leaders, economists). Frequently, problem owners have many other considerations to balance alongside ergonomics.

This means that if you want to implement ergonomics improvements, it is important to be able to analyse and discuss the trade-off between short-term demands of company leadership and the long-term benefits of ergonomics – in the language of cost-benefit analyses.

There is a paradox in the "economics of ergonomics"; when you choose to invest in good ergonomics proactively, it is hard to know exactly how much unnecessary cost has been avoided. This can sometimes make it challenging to convince management who are reluctant to make ergonomics investment. On the other hand, waiting to address bad work environments and work design until the workforce has been injured can spin off into a chain of costly effects (assembly errors, quality deficiency, sick leave, rehabilitation, compensation, costs for new recruitment, training of new staff and quality/speed deficiencies while new staff are under training, etc.)

For this reason, gaining knowledge from case studies and company records is a good way to develop arguments showing how the costs of bad ergonomics can propagate. From another angle, there are many case studies showing that improved ergonomics can improve safety, productivity, efficiency and quality, which all lead to profitability. Your ability to reason in these terms can greatly leverage your success in convincing other stakeholders and implementing workplace improvements in general, not just ergonomics.

WHICH ROLES BENEFIT FROM THIS KNOWLEDGE?

Any role that takes part in discussions of whether to invest in changes to the workplace can benefit from understanding the short- and long-term mechanisms of targeting ergonomics problems to avoid costly repercussions later. Typically, not many people who partake in such a discussion at companies are necessarily educated in seeing those connections, and therefore any engineer with an understanding for the economical benefits of good ergonomics can present a sound business case that shows what risks may end up costing much more than the initial short-term calculation may show. The *manager/leader* and *purchaser* who understand the difference between short- and long-term elimination of costly ergonomic risks and inefficiencies may both still be rare, but we hope to inspire more people in positions of leadership and economic responsibility to leverage this knowledge – particularly the less obvious investments into cognitive and psychosocial improvements. The *system performance improver* and *work environment/safety specialist* obviously benefit in their business case arguments from using examples from previous successful interventions, and highlighting which short- and long-term benefits may positively impact the work, the workers and the company. The *sustainability agent* would do well to add these economic perspectives to any discussion of how to make the workplace more sustainable from a combined social and economic perspective – for any intervention that makes good business sense has a better chance of making a lasting positive impact on operations.

11.1. Proactive or reactive approaches to ergonomics investments

Despite all the evidence that the design of a workplace and its associated tasks can trigger MSDs, causing sick leave and long-term illness for employees, many companies do very little to implement ergonomics principles in their business activities. Typically, companies only adopt a *reactive* approach when investing in ergonomics – they wait until the situation has become so bad that they have to react. This means that "quick and dirty" short-term solutions are implemented when complaints arise, but these solutions may not solve the root issue or provide lasting benefits. A reactive approach doesn't stop ergonomic issues from arising; rather, it means a number of people are "sacrificed" to the poor design of the workplace before anyone commits attention or resources to changing it.

In reality, the majority of ergonomics issues result from the design of the product and its associated assembly tasks, and so are actually already established in the design and planning phases, often years before production even begins. So adopting a *proactive* approach — where ergonomic considerations are planned in years ahead by designers, decision makers and production engineers — is a far superior approach. That will not only provide safe and healthy workplaces for employees, but is also likely to facilitate increased levels of productivity. There are many interconnected factors that influence production ergonomics, the majority of which are dictated by the design of the actual product itself. Figure 11.1 (from Munck-Ulfsfält, 1997) shows how all these factors affect the conditions of the assembler.

By adopting a proactive approach, it is possible to establish an assembly method with a minimal amount of ergonomics problems. As can be seen in Figure 11.2, the level of influence on ergonomics is highest at the start of the project before any design decisions have been finalized. The costs to make changes are also lowest at the start of the project, so it is most favourable from an economic standpoint to adopt a proactive approach.

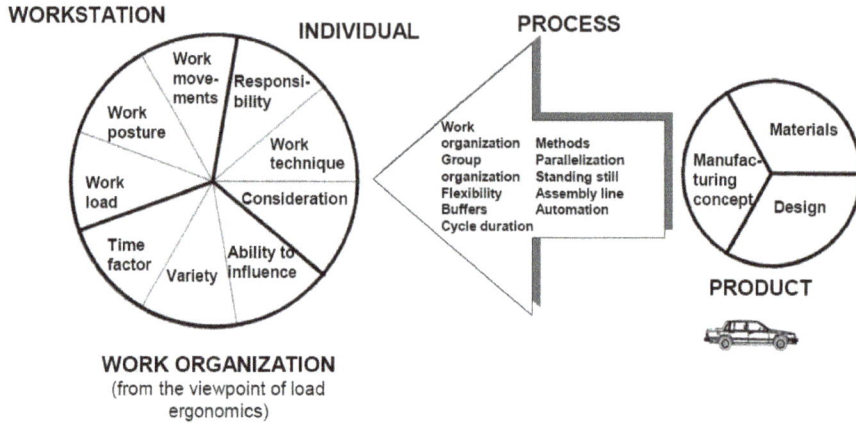

Figure 11.1: Holistic view of factors that affect production ergonomics (Munck-Ulfsfält, 1997).

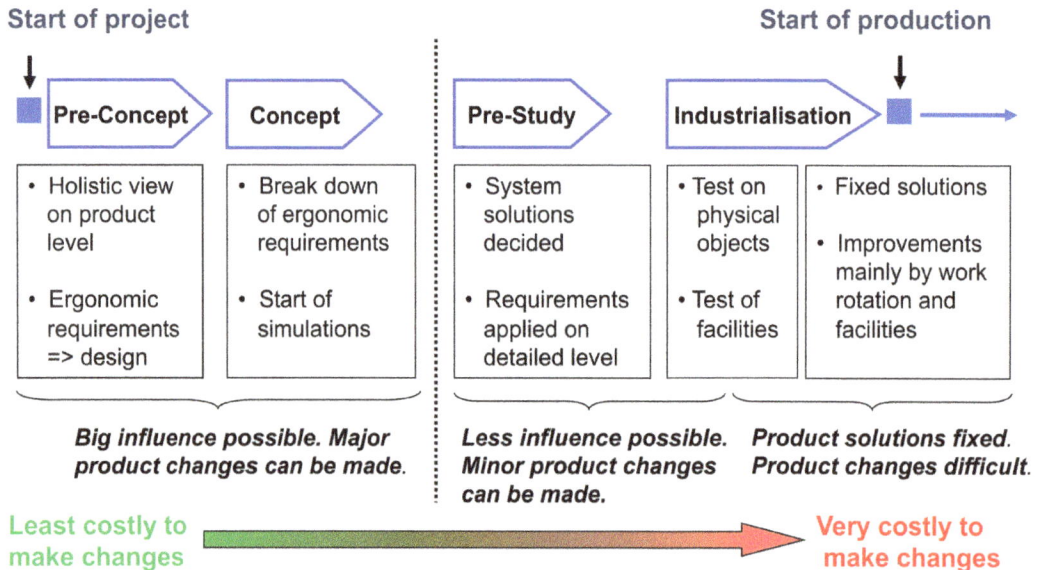

Figure 11.2: Level of influence and associated costs for good production ergonomics (Lämkull, Falck and Troedsson, 2007).

When trying to implement ergonomics changes, there is a need to speak a language of economics – something that many ergonomists and engineers often have limited experience of. If the awareness of negative consequences is low, it becomes necessary to quantify desirable changes in terms of costs. Education and training is also key – introducing engineers to ergonomics at university level is an important enabler for system improvements through ergonomics. Providing stakeholders and decision makers at all levels within the company (including workers) with training is also key in enhancing the workplace. The biggest scope for cost savings in companies comes from adopting a proactive approach to ergonomics, ensuring from the start of any project that harmful postures, high loads, poor tooling and excessive materials handling is avoided. If companies wait and take a reactive approach, injuries continue to happen and the ability to eliminate them will be limited and involve high costs. By contrast, a proactive approach enables modification to be made to the design and hazardous ergonomic conditions can be avoided.

Fundamentally, the lack of consideration of ergonomics by companies is due to economics and an uncertainty as to whether the costs associated with ergonomics really pay off in the long run. The costs associated with poor ergonomics aren't only associated with money and can be seen at three different levels:

- Costs to the individual
- Costs to the company
- Costs to society

11.2. Individual costs

From an individual perspective, there are a number of costs associated with a work environment that fails to take ergonomic considerations into account. The following list, adapted from Niu (2010 p. 748) exemplifies a few of the costs that burden individual workers:

- Pain and suffering due to injuries and occupational diseases (including Repetitive Strain Injuries (RSI), Cumulative Trauma Disorders (CTD) and repetitive motion injuries)
- Medical care costs
- Lost work time
- Lost future earning and fringe benefits
- Reduced job security and career advancement
- Lost home production and child care
- Home care costs provided by family members
- Adverse effects on family relations
- Lost sense of self-worth and identity
- Adverse effects on social and community relationships
- Adverse effects on recreational activities

Once a worker has an MSD or is experiencing pain on such a level that they can no longer carry out their job, the costs start to be counted. In many cases, workers are able to go back to work after a few days on sick leave. During prolonged periods of sick leave, however, individuals will no longer receive their full salary – rather, they are given sickness compensation, which is considerably lower. In the long term this will have significant impact on their personal economy, as their pension will also be affected.

There are also less measureable costs for the individual. While sick, opportunities for promotion and career advancements are significantly reduced; this can in turn affect the individual's job satisfaction and self-esteem. There is also a high degree of personal development and social interaction that comes with a job and losing these can have lasting psychological impacts on people. The combination of these costs in the form of lost money, time, competence and opportunities, can sometimes have a cumulative negative effect; injuries inhibiting a person's capacity to work can in some cases lead to cases of depression.

Ergonomics problems on an individual level are commonly solved (reactively) by medical staff using rehabilitation techniques. In some cases the root cause of the injury is identified, however this is not always the case, so without design changes the injury trigger could still remain a latent risk to other workers.

11.3. Company costs

A large number of costs resulting from poor ergonomics fall on the company. These costs can be categorized into the following areas:

- "Presenteeism"
- Sick leave
- Employee turnover
- Production losses
- Quality and business losses
- Legal costs

"Presenteeism"

Workers who have sustained an injury during their job can be split into two categories: those who are in pain but still manage to go to work, and those who are so injured that they can physically no longer work. Thus *presenteeism*, as opposed to absenteeism, is the state when workers ignore discomfort or pain and keep working. For workers that manage to work through the pain, costs still accrue to the company as the injured worker can no longer carry out their work to the same standard or speed, so the quality and productivity of their tasks will decrease. There may also be instances where they take a few hours off here and there to visit the hospital, meaning either the work won't be done or others will have to be paid to work overtime. Possible consequences of presenteeism include an increased risk of sudden injuries and accidents, lower product quality, slower pace of work, greater worker dissatisfaction and increased scrap rates.

Having unfit workers in the workplace can also affect morale as they start to resent their work; this can have a demotivational effect on other workers. In some cases, if possible, the production leader occasionally moves affected workers to so-called easier work tasks. However, these kinds of tasks are usually very limited, implying that this solution could not be offered to all who need it, nor for very long periods of time.

Sick leave, or absenteeism

The costs are even more significant for the company, when employees need to take prolonged periods of sick leave. Not only does the company need to continue to pay the sick individual's salary, but they may also need to pay others to work overtime, or in some cases have to hire subcontractors or recruit new personnel. When staff are fit enough to return to work, there may also be rehabilitation and retraining costs that the company will incur. There may also be less spare capacity to deal with emergencies that require extra staff.

Employee turnover

It is often said that people are the most valuable resource in a company and that their value increases with experience. However, poor work environments lead to a high turnover of personnel as people are either too injured to work or choose to leave the company as the job does not fulfil their needs and they fear that they will become sick if they stay too long. Hiring new staff to cover for absent individuals can be a costly and time-consuming process. Time, money and resources are spent advertising, interviewing, hiring and training new staff; there are also phase-out costs associated with employees who leave. It is very unlikely that new employees will work at the same rate as the sick experienced staff member they are replacing, so productivity rates are bound to slow initially. Time and attention is also taken up from other experienced workers who have to coach and support the new employee.

Production losses

Production losses frequently occur as a result of poorly designed workplaces and absenteeism. Productivity rates drop with the introduction of new employees and the increase of errors. Significant time may also be spent investigating injuries or accidents, reducing the production capacity of certain parts of the assembly line further.

Quality and business losses

A number of quality-related issues stemming from poorly designed workplaces have costs associated at a company level. A number of errors may occur when workers are in pain, fatigued, forced to adopt poor postures, unmotivated or bored. The introduction of new inexperienced employees or workers from other departments being called in to cover a shift can also lead to errors. The loss of productivity resulting from absent staff means meeting deadlines becomes harder, resulting in increased stress levels as staff rush to meet demand, which can also result in errors. At best, these errors means some components have to be scrapped or time spent to modify them, and at worst the error could go unnoticed meaning the product has to be recalled after it has already gone to market and reached the customer. Quality losses mean that scrap costs increase, as well as large sums of money being spent on recalls and warranties. There are also costs associated with the loss of the company's public image, their reputation will suffer and credibility will be lost, which will not only affect profits but can also make it harder to attract new (much-needed) employees. Such scenarios mean that focus, time and resources are spent by managers conducting accident and error investigations, rescheduling tasks and supervising workers. All of these cause reduced managerial focus as all their time is spent resolving urgent issues and keeping the company afloat, rather than planning for the long term.

Legal costs

In some cases workers press charges and take companies to court over the poor working environment they were subjected to, resulting in companies paying substantial fines as well as their corporate image being damaged. In general, a reactive culture of workplace health and safety may lead to higher insurance and compensation claim costs.

11.4. Societal costs

Bad workplace design triggering the onset of injuries also presents costs to society as a whole. According to research by Leigh (2011), medical and indirect costs of occupational injuries in the USA amount to at least the same costs as cancer, and since worker's compensation programs cover less than 25% of those costs, the economic burden befalls society. According to statistics from the International Labor Organization (ILO), work-related injury and illness costs vary between 1.8 and 6.0% of GDP in country estimates, with the average being 4%, and if involuntary early retirement is counted into the economic burden, the percentage can rise to staggering levels, such as in Finland (up to 15% of GDP!) according to Takala et al. (2014). The number of people needing medical care as a result of badly designed workplaces is also an issue for society, since so many hospital resources are taken up. For example, in 2007 the Swedish Social Insurance Agency, *Försäkringskassan,* paid over 99 billion SEK in benefits to individuals (Försäkringskassan, 2008). Using legal sanctions, this cost is often passed on onto companies, so instead of solving the root issue, the blame is just shifted.

11.5. Solving the problem

Despite the fact that the costs connected to poor ergonomics are vast and affect a number of people at different organisational levels, gaining approval from stakeholders for changes can be a challenging task. Given the multitude of different investment options across different sectors in companies, gaining approval from top-level management and sufficient funding to carry out projects can be a battle. With limited resources, companies have to prioritize needs and tend to invest in the most profitable venture. Given that those specializing in ergonomics rarely have the power to make the final decision when it comes to finance, they need to persuade investors of the economical value before they can carry out their job as an ergonomist. This can be difficult as the language of economics is very different to ergonomics or engineering. Everything needs to be discussed and quantified in terms of financial savings and benefits. Obtaining accurate figures to convey this information can be difficult as in reality, if an effective ergonomics program is implemented proactively, the costs of what could have been are never really fully known. While some of the cost benefits are obvious, such as reduced sick leave and less worker compensation pay-outs, there are also hidden costs like loss of productivity, employee turnover and quality issues.

To be fully equipped to implement ergonomic principles on a large scale in an industrial context, it is necessary to have:

- An awareness of the benefits gained by other companies.
- Tools and methods to aid in quantifying the benefits.
- Effective communication skills to convince the necessary stakeholders.
- Knowledge and the power to act and implement change.

11.6. Building awareness

The best way to raise awareness is through case studies that highlight how greater attention to ergonomics has brought about numerous benefits and cost savings in another business. The car industry is one sector that has been particularly strong in ensuring ergonomics considerations are made from the outset of projects. In 1999 a study at Ford Motor Company found that the hidden costs associated with bad ergonomics were three times worse than the more obvious costs, giving a combined total of $141 million as a result of bad ergonomics (Figure 11.3).

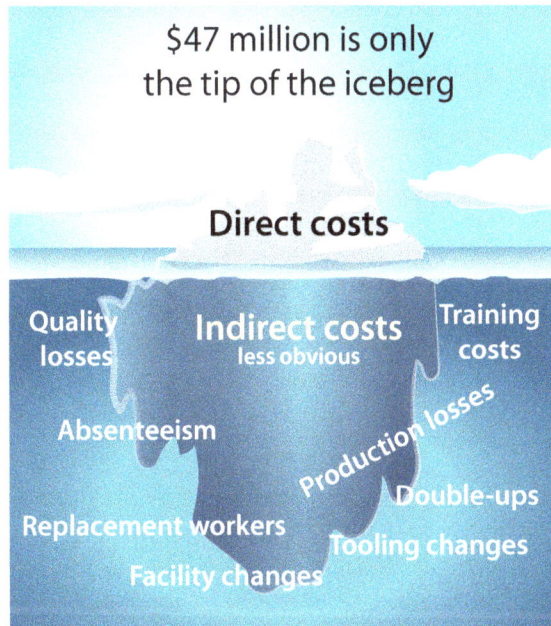

Figure 11.3: Direct and indirect costs of poor ergonomics that were once calculated at Ford Motor Company (adapted from Stephens, 1999).

Illustration by C. Berlin, based on Stephens (1999).

The costs associated with bad ergonomics were also noticed at Volvo Car Corporation; with estimates indicating that on average poor ergonomic work operations costs $170,000 annually (Falck, 2005). Another study at Volvo Car Corporation tracked the link between poor ergonomics and quality, in the form of errors. By following the assembly of 24,442 cars and monitoring the physical load levels of assembly tasks it was found that tasks with medium or high level would result in a 3.5 times higher risk for quality deficiencies, leading to 8.5 times higher costs to manage the associated errors (Falck, Örtengren, & Högberg, 2010). Another study at Volvo highlighted the importance of taking a proactive approach to ergonomics. With action costs for errors in the factory discovered late in assembly costing 9.2 times more than those repairs discovered at the early stages. In addition action costs to correct quality errors that were only detected once the products had reached the market were a further 12.2 times more expensive to correct compared to actions taken in the factory (Falck, Rosenqvist, 2014). Such benefits from improved ergonomics aren't limited to the car industry; research by Hendrick (1996) identified 25 cases across numerous different industries (ranging from the forestry industry to food service stands at a baseball stadium) where the implementation of ergonomics programs provided benefits.

Another effective way to increase awareness throughout a company is through ergonomics training programs. Equipping workers at all levels (including production technicians, manufacturing engineers, design engineers, production leaders and team leaders among others) with knowledge about ergonomics and how poorly designed workplaces can be improved, highlights that ergonomics is everyone's responsibility. Educating decision makers, project leaders and those with the power to make proactive production changes is urgent in prompting change. While empowering workers to take control of their workplace means that issues will be identified and reported earlier so modifications can be made before it's too late. Training also provides significant benefits at an organizational ergonomics level, as workers are more likely to take responsibility for their work and look out for colleagues, which increases morale creating a better working atmosphere.

11.7. Cost calculations

Cost calculations are the most effective way of convincing investors of the value of implementing ergonomics programs. Chances are that if an ergonomics improvement is seen as an improvement opportunity with measureable gains (Budnick, 2012) rather than just an investment cost, the more likely that the investment will be made.

Given the diverse and international nature of manufacturing industry, with bases of cost sometimes being very specific to the rules and regulations of specific countries, it is very difficult to use one standard tool for every situation or company. Numerous different attributes need to be taken into consideration when determining costs, so obtaining all the necessary data to make an accurate calculation may produce a need for very specific calculation models that are especially adapted to the country and situation in question. Nevertheless, we will try to discuss some general principles for cost calculations.

In general, demonstrating the value of improved ergonomics involves calculating the return on investment (ROI), a very basic metric that can be expressed by the following equation (often expressed as a percentage):

$$ROI = \frac{\text{Gains from investment} - \text{Costs of investment}}{\text{Costs of investment}} \times 100$$

Finding sources for ergonomics gains involves creative consideration of how a solution may positively impact the following system performance aspects (according to Budnick, 2012):

- productivity gains
- quality gains
- injury Prevention
- injury Management
- absenteeism reduction
- employee retention
- enhanced customer experience

It may be wise to focus on productivity and quality gains in particular, since most companies already measure and base a lot of decisions on those two metrics. In general, gains can be counted from two different perspectives when motivating an ergonomics investment: 1) an avoidance of losses, such as eliminating sick leave or scrap costs, or 2) increased revenue, such as increased output per time unit, higher quality, etc.

On a general level, demonstrating the value of investing in workplace ergonomics is about clearly demonstrating the balance between costs incurred by poor ergonomics, the cost of implementing a solution, and – most importantly – the economic returns that justify the investment into improving the workplace. You have to 1) determine the costs of losses and inefficiencies, 2) the costs of implementing the improvement, 3) the gains that result from the improvement, 4) the *time span* or *amount of products* that will measure the point at which the investment costs are compensated for by the improvement, i.e. the *break-even* point, and 5) the projected gains that will continue once the break-even point has been passed.

A convincing cost calculation will be very specific in detailing the costs of risks, time losses, sick leave costs, tool inefficiencies, materials scrap costs, etc. Sometimes, proposing a new solution becomes an exercise in not just motivating how the problems are going to be eliminated, but also demonstrating how many more additional gains can be achieved with a new solution. In other words, a bit of extra creativity goes a long way towards persuading the stakeholder with purchasing power. For instance, implementing a machine to do a previously dangerous or strenuous task may not just decrease sick leave costs, but also decrease the amount of time needed to make a product and the uniformity of the products, resulting in a higher output at a better level of quality, which can in turn increase sales profits.

The following case study (based on an article by Johrén, 2001) shows an example of the many factors that can contribute to improvement opportunities for ergonomics that can be presented as gains.

CASE STUDY

An electricity company was experiencing a number of ergonomics issues and was considering investing in sky lift equipment to reduce the load and injury risks on workers. To determine if the investment would pay off the following calculation was made (Johansson, Johrén, 2001).

Issues

- High absenteeism, 12 days per employee
- 20% have chronic back/joint problems
- 25% of the absence due to back problems
- Too strenuous for some female employees
- Strenuous for employees over 50 years of age
- Strenuous for employees with back/joint problems
- Risky working task, especially if the poles are rotten

Annual costs for skylifts	
Cost for one sky lift	150 000 SEK
Economic life length, 8 years	
Rest value, 10%	15 000 SEK
Interest rate, 15%	
Yearly cost	32300 SEK
Service & maintenance	7700 SEK
Sum	40 000 SEK
Annual cost for 10 sky lifts	400 000 SEK
Time gains at assembly with sky lift	
Time gain ½ h per pole	
100 poles assembled by working group & year	
Price 500 SEK/h (debiting price)	
Time gain for 1 sky lift?	
50 hours × 500 SEK/h	25 000 SEK/sky lift & year
Time gain for 10 sky lift?	
10 × 25 000 SEK	250 000 SEK/year
Time-gain per year:	**250000 SEK/year**

Reduced sick leave due to less back problems

- Today: 12 days sick leave per employee/year
- 25% regarded to be related to back problems
- Absenteeism cost estimated to 300 SEK/h
- Assume this can be reduced by 5%
  How large is the reduction in sick leave days then?

12 days × 165 × 25% × 5% = > 25 days

Reduced sick leave due to fewer back problems

How much are the costs for that absence reduced?

$25 \times 8\,h \times 300\,SEK/h = > 60\,000\,SEK/year$

Reduced sick leave cost: 60 000 SEK/year

Reduced absence due to fewer work accidents

Absence due to work accidents during the last year for employees with this task was 400 days; assume that the accidents can be reduced by 10%.

How much is this reduction in days?

Reduction: 40 days

How much are the costs for that absence reduced?

$40 \times 8\,h \times 300\,SEK/h = > 96\,000\,SEK/year$

Reduced cost due to fewer accidents: 96 000 SEK/year

Better use of the employees

Assume that people over 50 years of age and those with back and joint problems can work with a 2% productivity increase (20% of these 165 employees), with the total working hours per year = 1500.

How much is the revenue increased due to better use of the employees?

Revenue due to better use of employees: 495 000 SEK/year

(In addition: better job satisfaction, equality and customer satisfaction increase)

Total gain /revenue	
Gains from time savings at assembly	250 000 SEK/year
Reduced sick leave	60 000 SEK/year
Reduced absence due to fewer accidents	96 000 SEK/year
Better use of the employees	495 000 SEK/year
Total revenue	901 000 SEK/year
Profit	501 000 SEK/year

Sensitivity Analysis

Given that some assumptions were made in this calculation to prove that a profit would still be generated with more conservative figures a sensitivity analysis was also carried out.

- If sick leave reduction only is 2.5% (instead of 5%)
- If the accidents only are reduced by 5% (not 10%)
- If the use of employees with MSDs only increases with 1% (instead of 2%)

Costs	400 000 SEK
Revenues = 250 000 + 30 000 + 48 000 + 248 000	**576 000 SEK**

11.8. Case studies of ergonomics interventions

As part of convincing other stakeholders that an ergonomics-related design change will pay off, it may help to demonstrate examples of other cases where an intervention has been proven to have a positive economic impact. A number of case study examples have been collected in ergonomics literature to document how the removal of a health and safety risk resulted in several other gains as well, such as increased efficiency, speed, fewer accidents and wasteful mistakes, etc. Although calculation methods vary a great deal and the aspects taken into account are different, there are a number of examples of successful implementations of everything from new personnel routines to safety gear to weight handling equipment. Hal Hendrick (1996) describes several such cases in an article titled "Good Ergonomics is Good Economics", and several websites have compiled case studies to prove that by and large, Hendrick's catchphrase still rings true. The *Ergonomics Cost-Benefit Case Study Collection* provided by the Puget Sound Chapter of the Human Factors and Ergonomics Society (PSHFES, 2012; Goggins et al., 2008), and the case study collections housed on the websites of the United States Department of Labor (OSHA, 2016), the UK's Health and Safety Executive (2016) or the Canadian Centre for Occupational Health and Safety (CCOHS, 2015) all provide an array of examples of how ergonomics interventions played out in different work sectors.

11.9. Tools and calculation methods

A number of methods for ergonomics return on investment calculations exist, of which some are tools and others are calculation principles. While a number of tools do exist that relate business benefits in terms of cost to ergonomics, at present the lack of awareness and understanding coupled with the tools insufficient level of detail creates a barrier to their successful implementation.

The following sections list a variety of available cost calculation methods and tools, some of which are available online. The selection is mainly based on Rose and Orrenius (2006), but the list has been curated to include source materials in English and (mostly) publicly available tools (accessible via the provided links).

Calculate costs for sick leave absence

Some social security services may provide services for employers and employees to calculate the costs of individual sick-leave. For example, the Swedish Social Insurance Agency *Försäkringskassan* provides an online cost calculator to calculate individual costs for work absences due to ill health or injury. This calculator is adapted to Swedish social security regulations and compensation rates. The calculator is available online (Försäkringskassan, 2017).

SCA and MAWRIC

SCA(Statistically Based Cost Analysis Method) and MAWRIC (Method to Analyse Work related Risks, Improve work environment and estimate total Cost) are both developed by Rose (2001). SCA is used to gain an overview of the costs at group or company level for company- or sector-specific MSDs. MAWRIC is used to identify and assess MSD risks caused by specific tasks or occupations, and to suggest improvement.

Data required to use the methods:

- company or sector statistics of injuries and sick leave
- estimated productivity losses due to presenteeism
- risk assessments
- estimation of metrics after improvement
- data for costs and earnings

ROHSEI (Return on Health, Safety and Environmental Investments)

ROHSEI (Linhard, 2005) is a tool intended for team use, allowing typical financial metrics to be applied to health and safety improvements. It is described as a four-step process as follows:

- Understand the opportunity or challenge.
- Identify and explore alternative solutions.
- Gather data and conduct analysis.
- Make a recommendation.

Output metrics include net present value, return on investment, internal rate of return, and discounted payback period. (Also available as a commercial software through ORC Networks, 2011.)

Net cost model for workplace intervention

In this questionnaire-based method, described in Lahiri et al. (2005a and 2005b) and Lahiri (2005), net costs and net gains are calculated at company level for proposed ergonomics interventions targeted at decreasing the occurrence of MSDs and work disabilities (e.g. hearing loss). The net costs are calculated over a year and the method can calculate the investment's payback time. The method is available in the appendix of Lahiri et al. (2005a) or at http://faculty.uml.edu/slahiri/supriyajan28-website.doc (Lahiri, 2005).

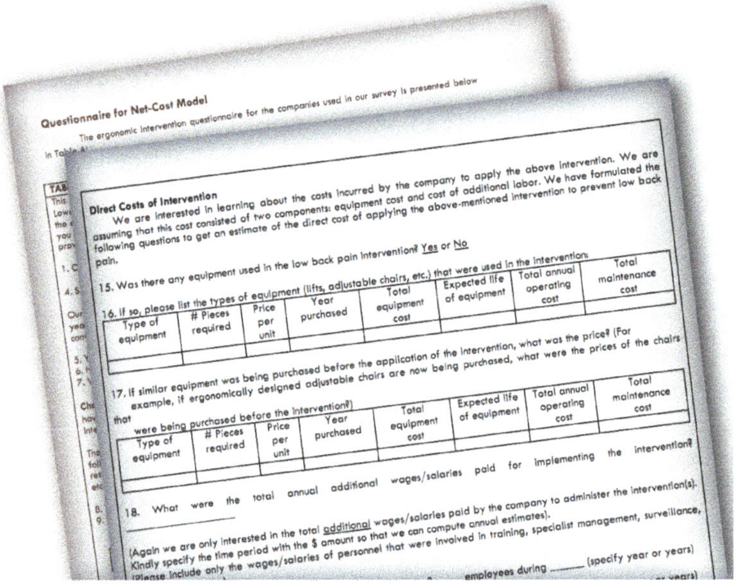

Figure 11.4: Illustration of the Net Cost Model questionnaire.

Image by C. Berlin, based on Lahiri (2005) and Lahiri et al. (2005a, 200b, 2005c and 2005d), with support from the WHO.

The Productivity Assessment Tool

Developed by Oxenburgh and Marlow (2005), this tool (Also known as the ProductAbility Tool in its software version) is a calculation tool that considers the following aspects (adapted from Oxenburgh and Marlow, 2005 p. 211):

Data concerning employees	Number of employees, their working time and wages, overtime, and productivity
Data concerning the workplace	Supervisory costs, recruitment, insurance, and other general overheads, maintenance, waste, and energy use, as applicable
The intervention	In the test cases the costs, or estimated costs, for the intervention
The reports	Cost-benefit analysis calculations and reports of the workplace and the employees

Washington State ergonomics cost-benefit calculator

Developed by the Puget Sound Chapter of the Human Factors and Ergonomics Society (PSHFES), this calculator (PSHFES, 2012), which is based on a review of 250 cost-benefit analysis cases for ergonomics investments (Goggins et al., 2008) is available as an excel file with pre-specified fields for specifying costs for work-related MSDs.

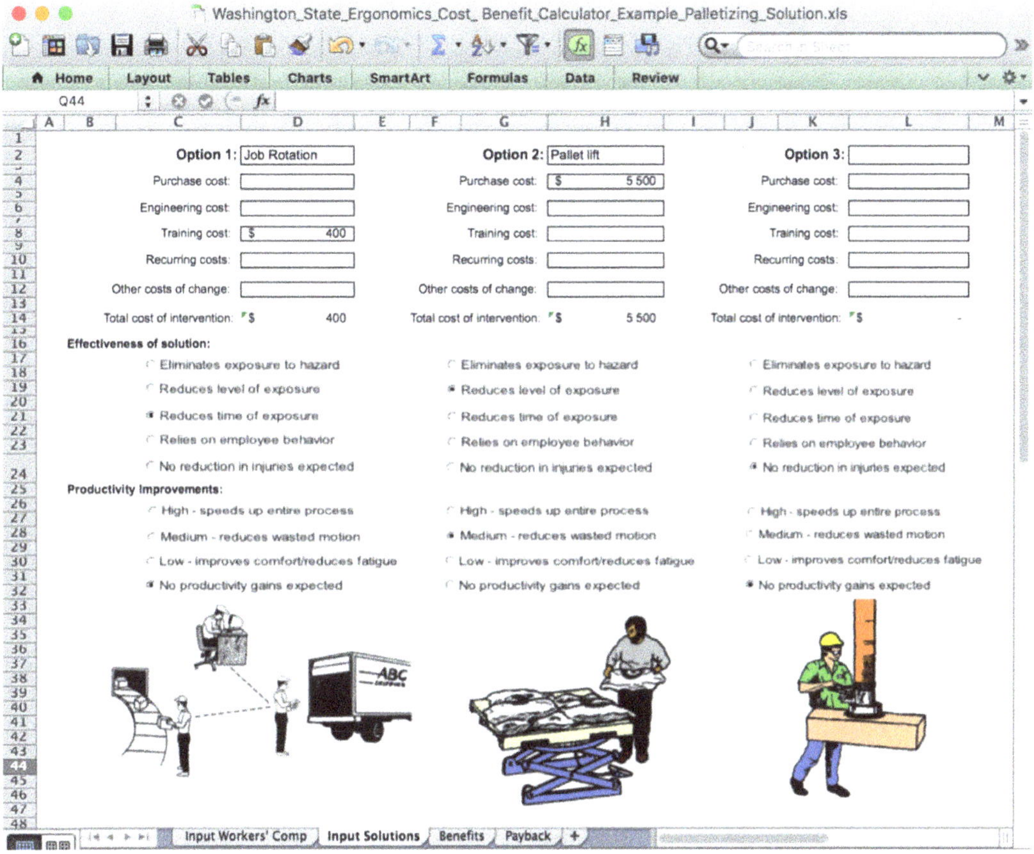

Figure 11.5: Screenshot excerpt of the Washington State Ergonomics Cost-Benefit Calculator (PSHFES, 2012). The calculator is available at http://www.pshfes.org/cost-calculator.

11.10. Special case: a model for calculation of poor assembly ergonomics costs

With a focus on the *product quality* consequences of poor assembly ergonomics in the automotive industry, Falck and Rosenqvist (2014) have developed a product-focused calculation method that is meant for "engineers and stakeholders in the design or redesign of manual assembly solutions" (p. 140). Based on the assembly of 47,061 cars, the method calculates costs based on a found correlation between product quality errors (which led to costs in the form of scrap, blocking of production, errors, recall and repair of products, staff costs for additional efforts, customer dissatisfaction and brand devaluing) and tasks rated as having poor ergonomics.

The calculation itself looks like this (adapted from Falck and Rosenqvist, 2014 p. 144):

Costs of poor assembly ergonomics, related to products

C = total costs of manual assembly errors   **W** = labour cost/time unit

Costs (C):

C_{scrap} = scrap cost per item C_{fb} = cost of errors of factory blocked cars

C_{fcomp} = cost of errors of factory complete cars C_{rec} = cost for recall/repair of cars distributed to the customers

C_{effort} = cost of staff/time unit in additional efforts, C_{bw} = cost for bad will (lost brand image and
e.g. meetings, controls, expanded staffing, etc.   customer's dissatisfaction)

$WRSL$ = cost of work-related sick leave and rehabilitation

Number of errors (N):

N_{on} = number of quality errors online N_{off} = number of quality errors offline

N_{au} = number of audit quality remarks N_{yard} = number of cars in the yard awaiting repair

N_{scrap} = number of scrapped items N_{fb} = number of factory blocked cars

Number of extra staff (N):

N_{effort} = number of people involved in additional efforts

Action time (T)

Ta_{on} = action time online Ta_{off} = action time offline

T_{ty} = transportation time for cars in the yard to/from work shop

$$C = W\,(N_{on} \times Ta_{on} + N_{off} \times Ta_{off} + N_{au} \times Ta_{off} + N_{yard} \times T_{ty}) + N_{scrap} \times C_{scrap} + C_{fb} \times N_{fb} + WRSL$$
$$+ C_{fcomp} + C_{rec} + C_{effort} \times N_{effort} + C_{bw}$$

The calculations have components that are obviously related to automotive production (and re-work) conditions in particular, but the coverage of the indicators provide a very well-specified guide to what costs to look for.

11.11. Convincing the necessary people

When trying to implement ergonomics programs it is necessary to convince and communicate with a number of stakeholders throughout the company, especially the decision makers who have the majority of power:

- investors
- managers
- operators
- logisticians
- sub-contractors, suppliers
- health and safety group
- unions

For the majority of stakeholders, discussions focused around costs is the most convincing way to present the case. There are three different types of cost, all of which should be used when presenting the case (Zandin, 2001):

- historical costs
- projected savings
- actual cost savings

Historical costs relate to the issues that have already occurred due to poor ergonomics, such as time and money lost due to prolonged periods of sick leave. Introducing the problem with such costs alongside some relevant statistics and complaints from workers provides a credible foundation to present your case. It enables stakeholders to understand that the current approach is problematic and should be modified. Projected savings can be estimated using various cost calculation tools based on knowledge of historical costs and assumptions. Providing stakeholders with projected savings can be particularly beneficial during the prioritization of which new projects to take on and the decision-making process. Following the ergonomic intervention it is then possible to identify actual cost savings. These savings should be compared with the projected savings so assumptions can be validated and modified if necessary. These savings can be used as case studies for future ergonomics interventions both within the company and externally, helping to spread awareness of the economic value of ergonomic interventions.

In addition to presenting cost benefits, other potential improvements should also be highlighted. Such as improved working atmosphere with more motivated staff and improved company reputation, attracting a larger pool of perspective employees.

11.12. The power to implement change

Once the stakeholders have agreed to implement an ergonomics project, it is important to carry out the change in a logical and holistic way to maximize gains and clearly demonstrate positive the effects of the change. The latter can be done by making sure to present the change in an attractive way so that all levels of involved stakeholders, from management to worker, are aware of the original reason for the change process itself, as well as the short- and long-term effects of it. By utilizing the knowledge you have gained in ergonomics and work design, you can highlight different perspectives of how the change play out for different stakeholders, such as pointing out how the work is easier and more efficient for workers, stating to management which unnecessary losses have been avoided, etc. Making sure to record the impacts of the intervention help you build up your own library of "success stories" that can increase trust among other stakeholders in the improvement proposals that you suggest.

Study questions

Warm-up:

Q11.1) Name at least three individual-level costs of poor workplace ergonomics.

Q11.2) When a company loses an employee to sick leave, what are the potential "hidden" costs that can be incurred?

Q11.3) What industrial costs can result from poor assembly ergonomics?

Q11.4) When a company makes an investment in a workplace change targeting ergonomics, what are some examples of expected gains that the investment can be balanced against?

Look around you:

Q11.5) Find two ergonomics intervention case study descriptions (for example, you can find them by using the sources quoted in section 11.2.3) and compare – what were the expected economic gains, and how was the return on investment accounted for in the case study?

Q11.6) (Continuation of the previous question) Compare what kinds of problems were identified in the case studies as impacting economic performance, how the intervention was designed to address the problem, and what outcomes were measured to prove whether the intervention was successful in an ergonomic and economic sense.

Connect this knowledge to an improvement project

- If you are drafting a suggestion for a workplace design change or ergonomics intervention, find an example of a similar case study (preferably describing a similar intervention or implementation) and use it as an example of the proven benefits in another setting – this proof-of-concept may help you argue for the proposal being a good idea also in the context you are addressing.
- When calculating the impacts of an investment, include as many different economically related safety, productivity and efficiency aspects as you can think of. Think of both long- and short-term impacts on the operations of the workplace and/or the company's output capacity.
- Remember to emphasize impacts both on the individual, team and company levels if possible. Starting with a small implementation that proves successful may pave the way for larger investments in the future.
- Use a sensitivity analysis to calculate a more modest prognosis of the gains – this will help to convince sceptical stakeholders that the investment is not too optimistic.

Connection to other topics in this book:

- Using digital human modeling (Chapter 9) and model representations of the workplace to gain worker input and acceptance for changes (Chapter 6, section 6.5) are cost-effective ways to test alternative solutions without exposing workers to risk or wasting money, materials and time on solutions that are unsuccessful.

• A socially sustainable workplace (Chapter 13) tends to enable and encourage workers to stay at a workplace and over time develop skills and knowledge that make them even more valuable to the employer – knowing how to keep these employees safe and motivated is a long-term economic investment in any company's future.

Summary

• Research and several case studies show that a strong link exists between a good work environment and company profits.
• Poor ergonomics leads to a number of costs for individuals, society and companies.
• Adopt a language of economics when trying to convince stakeholders to invest in proactive ergonomics efforts.
• Utilizing a proactive approach considering ergonomics from the projects outset provides significant cost benefits compared with a reactive approach.
• Building awareness of the proven benefits of ergonomics programmes can be done through case studies and training.
• A number of tools and cost calculation methods can be used to quantify ergonomics benefits. Some are targeted at a particular aspect of cost, such as sick leave, product-related losses, payback time, etc.

11.13. References

Budnick, P. (2012) *Ergonomics ROI: How to Document Ergonomics-Related Improvements*. [Online]. Available from: https://ergoweb.com/ergonomics-roi-how-to-document-ergonomics-related-improvements-reprint/ [Accessed 27 July 2016].

CCOHS (2015). Workplace Health Case Studies. [Online] Available from: http://www.ccohs.ca/healthyworkplaces/employers/casestudies.html [Accessed 27 July 2016].

Falck, A. (2005). Good Ergonomics – Can we reckon the benefits? Svenskt Monteringsforum, Monteringsrådets konferens, Stockholm, 2005. (Conference Proceedings in Swedish).

Falck, A.-C., Örtengren, R. & Högberg, D. (2010). The impact of poor assembly ergonomics on product quality: A cost-benefit analysis in car manufacturing. *Human Factors and Ergonomics in Manufacturing & Service Industries,* 20(1): 24–41.

Falck, A.-C. & Rosenqvist M. (2014). A model for calculation of the costs of poor assembly ergonomics. *International Journal of Industrial Ergonomics,* 44: 140–147.

Försäkringskassan. (2008) *Social Insurance in Figures*. [Online]. Available from: http://www.forsakringskassan.se/wps/wcm/connect/48d349cf-3387-4085-aed8-6a56ee1959f6/socialforsakingen_i_siffror_2008_eng.pdf ?MOD=AJPERES [Accessed 18 January 2014].

Försäkringskassan. (2017). Calculate costs for sickness absence. [Online]. Available from: https://www.forsakringskassan.se/arbetsgivare/e-tjanster-for-arbetsgivare/berakna-kostnader-for-sjukfranvaro [Accessed 28 feb 2017]

Goggins, R. W., Spielholz, P. & Nothstein, G. L. (2008). Estimating the Effectiveness of Ergonomics Interventions Through Case Studies: Implications for Predictive Cost-Benefit Analysis. *Journal of Safety Research*, 39(3): 339–344. DOI: https://doi.org/10.1016/j.jsr.2007.12.006.

Johansson, U. & Johrén, A. (2001). *Personalekonomi idag*. Uppsala: Uppsala Publishing House (in Swedish).

Health and Safety Executive (2016). [Online] Available from: http://www.hse.gov.uk/humanfactors/resources/case-studies/ [Accessed 27 July 2016].

Hendrick, H. W. (1996). "Good Ergonomics is Good Economics." [Online]. Available from http://www.hfes.org/web/pubpages/goodergo.pdf [Accessed 27 July 2016].

Lahiri (2005). Questionnaire for Net-Cost Model. [Online] Available from: http://faculty.uml.edu/slahiri/supriyajan28-website.doc [Accessed 28 Nov 2015].

Lahiri, S., Gold, J. & Levenstein, C. (2005a), Net-cost model for workplace interventions. *Journal of safety research*, 36(3): 241–255.

Lahiri, S., Gold, J. & Levenstein, C. (2005b). The Cost Effectiveness of Occupational Health Interventions: Preventing Occupational Back Pain. *American Journal of Industrial Medicine*, 48: 515–529.

Lahiri, S. Gold, J. & Levenstein, C. (2005c). Estimation of Net-Costs for Prevention of Occupational Low Back Pain: Three Case Studies from the US. *American Journal of Industrial Medicine*, 48: 530–541. Also in: *Revista Brasileira de Saude Occupacional*, Vol. 31 No. 113, 2006.

Lahiri, S. Gold, J. & Levenstein, C. (2005d) Net-cost Model for Workplace Interventions. *Journal of Safety Research*, 36: 241–255. Available from http://www.who.int/occupational_health/topics/lahiri.pdf [Accessed 27 Dec 2016].

Leigh, J. P. (2011). Economic burden of occupational injury and illness in the United States. *Milbank Quarterly*, 89(4): 728–72.

Linhard, J. B. (2005). Understanding the return on health, safety and environmental investments. Journal of Safety Research – ECON proceedings 36: 257–260.

Lämkull, D., Falck, A.-C. & Troedsson, K. (2007). Proactive ergonomics and virtual ergonomics within Manufacturing Department at Volvo Car Corporation. In: Berlin, C & Bligård, L-O. (Eds.) *Proceedings of the 39th annual Nordic Ergonomic Society Conference October 1–3, Lysekil*, Sweden. p. 118.

Munck-Ulfsfält, U. (1997). Production ergonomics in car manufacturing — evaluation of a model to achieve a good ergonomics result in existing production and in alteration work. In proceedings, IEA, Tammerfors, Finland, 1997, Vol.1, pp. 229–23.

Oxenburgh, M. & Marlow, P. (2005). The Productivity Assessment Tool: Computer-based cost-benefit analysis model for the economic assessment of occupational health and safety interventions in the workplace. *Journal of Safety Research* – ECON proceedings 36: 209–214.

ORC Networks. (2011). Return on Health, Safety and Environmental Investments (ROHSEI). [Online] Available from: http://orc-dc.com/?q=node/821 [Accessed 28 Nov 2015].

OSHA. (2016). Case Studies. [Online] Available from: https://www.osha.gov/SLTC/ergonomics/case_studies.html [Accessed 27 July 2016].

PSHFES. (2011a). Cost-Benefit Analysis. The Puget Sound Chapter of the Human Factors and Ergonomics Society. [Online] Available from: http://www.pshfes.org/cost-calculator [Accessed 28 Nov 2015].

Rose, L. (2001). Models and Methods for Analysis and Improvement of Physical Work Environments. PhD Thesis, Dep. of Product and Productions Development, Div. of Human Factors Engineering, Chalmers University of Technology, Göteborg, SE.

Rose, L. & Orrenius, U. (2007). Beräkning av arbetsmiljöns ekonomiska effecter på företag och organisationer — En översikt av ett urval modeller och metoder. Arbete och Hälsa Report No. 2006:18 (in Swedish). The National Institute of Working Life, Stockholm, Sweden.

Stephens, A. (1999). Direct and indirect cost for poor ergonomics. Ford Motor Company, Dearborn, USA.

Takala, J., Hämäläinen, P., Saarela, K. L., Yun, L. Y., Manickam, K., Jin, T. W., Heng, P., Tjong, C., Kheng, L. G., Lim, S. & Lin, G. S. (2014). Global Estimates of the Burden of Injury and Illness at Work in 2012. *Journal of Occupational Environmental Hygiene* 11(5): 326–337.

Zandin, K. B. (2001) *Maynard's Industrial Engineering Handbook*, 5th Edition. New York, McGraw Hill. [Online] Available from: http://accessengineeringlibrary.com/browse/maynards-industrial-engineering-handbook-fifth-edition/p2000a1fc99706.9001 [Accessed 16 January 2014].

CHAPTER 12

Work Environmental Factors

Image reproduced with permission from SasinTipchai/ Shutterstock.com. All rights reserved.

THIS CHAPTER PROVIDES:

- A brief introduction to the occupational hygiene discipline.
- Descriptions of environmental aspects in physics terms.
- Descriptions of how environmental aspects affect the human physiologically, cognitively and psychologically.
- Descriptions of how environmental aspects can be measured.

How to cite this book chapter:
Berlin, C and Adams C 2017 *Basic Anatomy and Physiology.* Pp. 213–240. London: Ubiquity Press. DOI: https://doi.org/10.5334/bbe.l. License: CC-BY 4.0

WHY DO I NEED TO KNOW THIS AS AN ENGINEER?

As you learned in Chapter 3, the human body has certain responses to physical loading. It might not be obvious at first, but many environmental aspects of our work environment can cause mental, psychological and physical loading on the body, especially in extreme work environments. When the body is reacting to stimuli from the environment, or has to work while wearing protective clothing and gear, it often puts us in a slightly weaker position to take on physical and mental loading from the actual work tasks. As a result, environmental aspects can be anything from annoying to distracting to hazardous and completely debilitating. As engineers, we can use our basic understanding of physics and measurement to assess different environmental factors. Usually, we make a present state analysis of the environment and compare it to ideal measurements for human performance, in order to establish criteria for a change. The specific discipline of evaluating work environmental factors is also known as *occupational (or industrial) hygiene*, and has its own professional organizations worldwide. This discipline tends to cover many additional environmental factors, so in this book we focus primarily on five types of work environmental factors that can affect cognitive performance, physical loading on the body or health and safety in general: thermal climate, lighting, sound, vibration and radiation.

Creating a suitable balance of these factors is part of designing a well functioning workplace. Understanding these factors, the ways that the human body reacts to them, and how to use instruments, standards and methods to evaluate their suitability, are all essential skills in the workplace engineering toolbox.

WHICH ROLES BENEFIT FROM THIS KNOWLEDGE?

The *system performance improver* can use the knowledge in this chapter to ensure that the working conditions that are not directly part of workplace and equipment design may still be accounted for in the furnishing of work spaces, the planning of work tasks, limiting exposure to particular environments and by supplying protective gear. Many standards for making the environmental factors exist, some of which can be used as helpful design guidelines. The *work environment/safety specialist* benefits from being able to point out standardized limit values for exposure to specific environmental conditions. These established standards can be a big help in backing up arguments of how to protect workers when the work demands particular conditions. Also, understanding the direct impacts of environmental exposure on human physiology and cognition can help in the process of suggesting solutions for how to target risks. The *sustainability agent,*

who may be tasked with addressing environmental sustainability also, may benefit from understanding how certain energy-consuming work conditions (for example keeping a space heated or cooled) can be reasoned about in conjunction with human well-being goals – at best, this role can pursue solutions that have beneficial impacts on all sustainability aspects.

12.1. The human body in different environments

As you have learned before, the human body has abilities and limitations that need to be considered in the design of work environments. For example: in cold conditions, blood flow to the outer extremities is restricted, which may lead to shivering that can impair the sense of touch and hinder precision work. In a loud, noisy environment, cognitive resources are split between listening for information and hearing non-meaningful noise, which can impair concentration and cause psychological stress. In dim light, it is a common occurrence that people need to bend closer to see what they are doing, causing poor ergonomics. In a vibrating environment, internal muscular loading increases because the body is reacting to the small forces propagating throughout its tissues, leading to fatigue. All of these examples illustrate that the physical work environment has a great influence on human well-being and system performance.

To design the optimal conditions to perform work, we need to be aware of the "comfort zones" in which the human body can best perform both mentally and physically, without wasting resources on extra loading from the environment. For this reason, knowledge of measures, tools and ideal limit values are of great value when designing a work environment.

12.2. Occupational (or Industrial) Hygiene

Since workplace improvement is not solely the concern of ergonomists, there is also a well-known discipline in its own right that focuses on work environmental factors, and to some extent its scope overlaps greatly with that of this book. This discipline is known as Occupational Hygiene (or in North America, Industrial Hygiene). We offer a quick introduction here to orient our reader to the discipline. The International Occupational Hygiene Association (IOHA) defines occupational hygiene as:

"the discipline of anticipating, recognizing, evaluating and controlling *health* hazards in the working environment with the objective of protecting worker health and well-being and safeguarding the community at large."

(IOHA, n.d)

Occupational/industrial hygienists concern themselves with many of the same aspects that ergonomists and production engineers do, but may come from a different educational background (quite often in engineering, chemistry, physics, or a biological or physical science), which places a more pro-nounced emphasis on assessing work-environmental concerns with a scientific measure-and-control approach. Some examples include:

- Indoor air quality, air contaminants
- Chemical exposure hazards, e.g. lead
- Emergency response planning and community right-to-know
- Occupational diseases (AIDS in the workplace, tuberculosis, silicosis)
- Biological hazards, e.g. bacteria, viruses, fungi
- Potentially hazardous agents such as asbestos, pesticides, and radon gas
- Ergonomic hazards, cumulative trauma disorders (repetitive stress injuries, carpal tunnel syndrome)
- Radiation (electromagnetic fields, microwaves)
- Reproductive health hazards in the workplace
- Exposure limits to chemical and physical agents
- Detection and control of potential occupational hazards such as noise, radiation, and illumination
- Hazardous waste management

(OSHA, n.d.; American Industrial Hygiene Association, 2016)

The discipline also features a well-accepted "Hierarchy of Controls" (NIOSH, 2015) according to the American National Standards institute (2005), which describes in which order any found health hazards are to be controlled. Primarily (for best control effectiveness and business value), the hazard should be eliminated, but if this is not possible, the order of preferable interventions is substitution (replacing the hazard); engineering controls (isolating workers from the hazard); administrative controls (changing how people work); and finally as a last resort, providing workers personal protective equipment. Figure 12.1 shows this hierarchy. This prioritization approach is used by occupational/industrial hygienists for all kinds of identified workplace hazards, including physical ergonomics and psychosocial aspects.

In this book we will not elaborate on all the physical and chemical hazards of industrial workplaces, as the subject is extensive in its own right and may vary in scope depending on the exact industrial sector being studied – for example, some work environments that are dusty or humid may have air quality as a major concern, while others may involve continuous exposure to chemicals, sprays, etc., and may require other approaches. However, there are many knowledge resources to turn to in the form of associations and professional organizations concerning themselves with occupational/indus-trial hygiene, e.g. the Occupational Safety and Health Administration (OSHA) of the United States Department of Labor; the International Occupational Hygiene Association (IOHA); the American Industrial Hygiene Association (AIHA), and many more. These organizations provide training, con-ferences, knowledge resources and a community of science and practice regarding workplace hazard identification. In the following sections, we will focus mainly on the work environmental factors known as *physical hazards* (OSHA, n.d.).

12.3. Thermal climate

Thermal climate is more than just temperature. The predominant impression of climate usually has to do with feeling too hot, too cold, or just right (which often corresponds to not noticing the climate at

BEST

ELIMINATION
Design it out

SUBSTITUTION
Use something else

ENGINEERING CONTROLS
Isolation and guarding

ADMINISTRATIVE CONTROLS
Training and work scheduling

PERSONAL PROTECTIVE EQUIPMENT
Last resort

BEST

Control effectiveness

Business value

Figure 12.1: The Hierarchy of Controls against occupational hazards; image from NIOSH (2013 p. 48). Image reproduced with permission from CDC/NIOSH. Image is in the public domain via www.cdc.gov.

all). The ideal thermal climate is known as comfort climate, which is the often-unnoticed psychophysical state in which humans experience satisfaction with the climate, and thereby the best conditions for work. However, the human experience of temperature is affected by other factors, like air speed, humidity and our ability to exchange heat with the environment. Therefore, climate is a complex environmental factor that can be assessed by studying the following climate parameters: heat, cold, comfort climate, heat exchange, clothes and insulation.

Thermal balance

A healthy human body automatically regulates thermal exchange with its surroundings in order to keep the temperature of the blood constant. If the blood temperature changes, the body is rapidly put into a state of discomfort and illness, often resulting in compensation behaviours, or in worst cases, impairing mental or physical performance.

The balance of the body's heat exchange with the environment can be expressed in terms of power (energy consumed per time unit, i.e. Joules/second = Watt). The balance in the following equation (Bohgard, 2009 p. 198) states that the *difference* (usually an increase) in power produced by the body and the power generated by the mechanical work must be balanced by a number of heat/energy losses of different kinds:

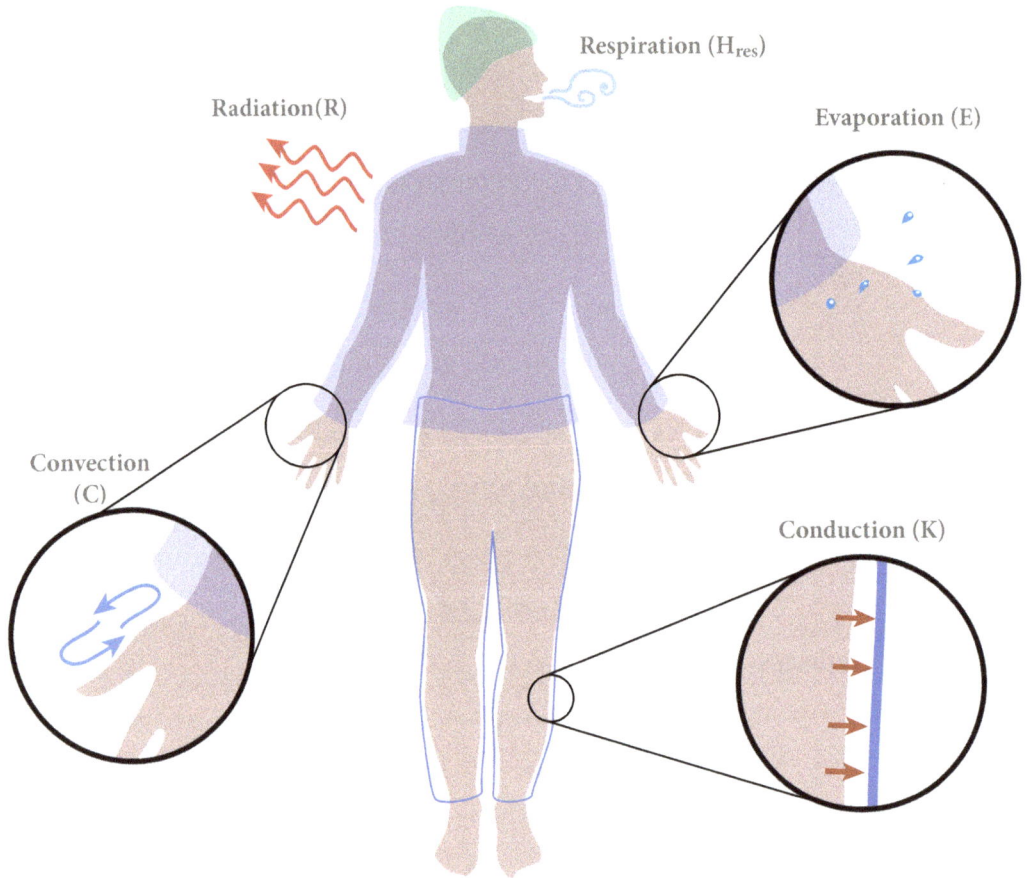

Figure 12.2: The different components of thermal power loss. Due to the thermal inertia of most clothing materials, conduction is presumably negligible regarding the contact of the clothing to the skin.

Illustration by C. Berlin.

$$M - W = R + C + K + E + H_{res}$$

Where:
  M = power produced in the body
  W = mechanical power
  R = power loss by heat radiation
  C = power loss by convection
  K = power loss by conduction
  E = power loss by evaporation from body surface
  H_{res} = (as in <u>res</u>piration) power loss by evaporation from airways

When the above equation is not in equilibrium, the result is a rise or fall of the body's temperature, often leading to discomfort and possibly higher injury risk and performance impairment. Symptoms of this imbalance often come in the form of complaints or physical manifestations, e.g. irritation, dryness, rashes, tiredness, headaches or muscular tension.

In different external temperatures, the components of heat exchange (Fig. 11.1) are proportioned differently. For example, in hot environments, it is likely that much heat transfer will be due to evaporation (sweating) and respiratory evaporation, while in a cold climate, the dominant components will instead be radiation and convection (air circulation, absorbed by clothing). The body compensates for extreme climates by regulating its own production of heat; for example, this is why we shiver in cold temperatures, activating heat in our muscles.

For reference, Table 12.1 shows the power outputs for a "standard person" measuring 175 cm (corresponding to a surface area of 1.84 m²) and weighing 70 kg:

Table 12.1: Metabolic rate (power output over surface area) for different activities (adapted from Bohgard, 2009 p.197) for a person measuring 175 cm (corresponding to a surface area of 1.84 m²) and weighing 70 kg.

Activity	Metabolic rate (W/m²) (based on ISO 8996 and ISO/TS 16976-1)
Rest	65
Light work	100
Moderately heavy work	165
Heavy work	230
Maximum	600

12.4. Thermal exposure risks

One thing worth mentioning is that a hot or cold environment also changes the temperature of materials, which can add to the physical and mental stress of performing assembly work since the hands are extra sensitive to heat and cold. This is especially true for metals, both in the form of tools and product material.

In very warm environments, exposure of the skin to a hot surface for certain duration of time may result in a burn, which is a painful, irreversible tissue injury. Depending on the temperature of the material that the skin is exposed to, we can tolerate different time durations of exposure up to the threshold of pain or the threshold of injury. The reason for the "delay" of the burn is due to thermal inertia, a property that describes the speed of the heat transfer between the material and the skin, and is dependent on the kind of material, the size of the contact point, surface characteristics, conductivity and other factors. For example, the thermal inertia of a piece of metal allows a much more rapid heat transfer to the skin, compared to a similar-sized piece of wood at the same temperature. This phenomenon happens on both extremes of the scale; in very cold temperatures, exposure to very cold materials or air for certain duration of time can lead to irreversible tissue damages in the form of frostbite.

12.5. Heat

The sensation of heat is experienced when the body senses its surroundings (air, water or material objects, usually via the skin) as being warmer than the body itself. Table 12.2 shows how heat at different levels can have the following effects:

Table 12.2: Effects of heat at different intensities.

Warm, within comfort zone	• Increased peripheral blood flow, widened blood vessels • Skin temperature rises • Drop in muscle tension
Moderate heat, just slightly beyond comfort zone	• Sweating • Loss of fluids and salt • Tiredness • Increased hostility • Decreased performance and alertness • Increased risk for mistakes and errors • Increased risk for accidents
Extreme heat, discomfort	• Painful cramps • Impaired function of stomach and intestines • Heat regulation failure

12.6. Cold

In cold environments, the human body is especially sensitive to the external climate due to additional factors that may alter the sensation of temperature. Wind chill temperature (twc) captures this by taking consideration of the nominal temperature and the effects of wind velocity. Many times, air velocity can increase the local skin sensation of the air temperature being colder than its nominal measurement, which can lead to increased risk for severe cold-related damages (see Table 12.3) at shorter time exposures. (This also refers back to the previous description of burn risk as a function of exposure time to materials at different temperatures.)

Table 12.3: Effects of cold at different intensities.

Cool, within comfort zone	• Reduction of blood flow to skin, constricted blood vessels • Decrease in performance due to thick clothing
Moderate cold, just slightly beyond comfort zone	• Discomfort • Shivering • Decreased fine motor function • Decrease in sense of touch
Extreme cold, discomfort	• Disorientation • Apathy • Weaker breathing • Frostbite: oxygen shortage to a part of the body resulting in tissue damage

12.7. Assessing climate

We can measure thermal climate in a number of ways. In order to design work environments that have satisfactory thermal climate, we need to consider all the influencing factors that lead to a human response:

- air temperature
- radiation
- air humidity
- air velocity
- clothing
- activity

The International Organization for Standardization – ISO – specifies several procedures for assessing climate, some of which are specialized for hot or cold environments and specific industrial applications (Figure 12.3). This chapter will bring up two of these: one objective parameter measurement called Wet Bulb Globe Temperature (WBGT, ISO 7243) and one subjective measurement of thermal comfort, called Predicted Mean Value/ Predicted Percentage of Dissatisfied (ISO 7730:2005).

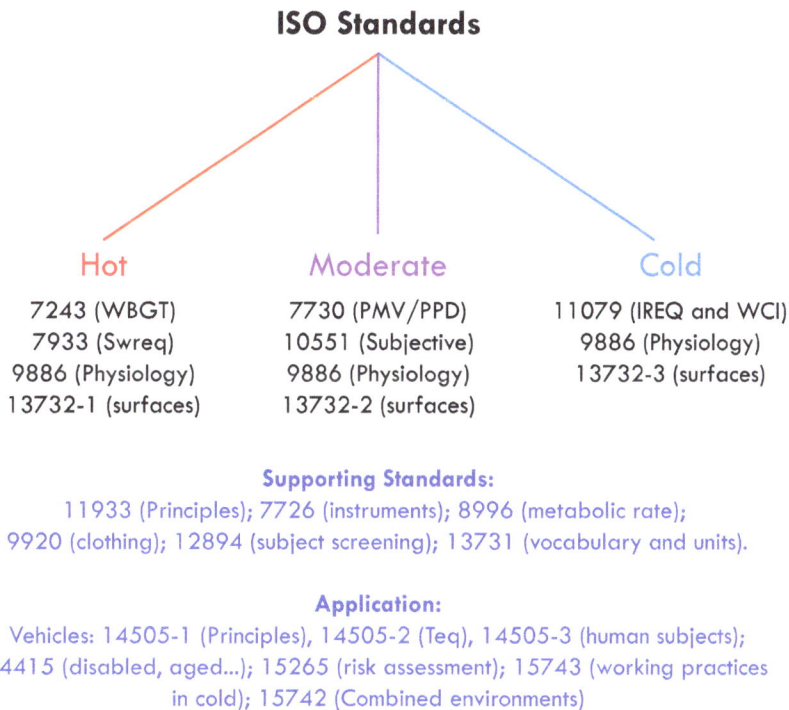

ISO Standards

Hot

7243 (WBGT)
7933 (Swreq)
9886 (Physiology)
13732-1 (surfaces)

Moderate

7730 (PMV/PPD)
10551 (Subjective)
9886 (Physiology)
13732-2 (surfaces)

Cold

11079 (IREQ and WCI)
9886 (Physiology)
13732-3 (surfaces)

Supporting Standards:
11933 (Principles); 7726 (instruments); 8996 (metabolic rate);
9920 (clothing); 12894 (subject screening); 13731 (vocabulary and units).

Application:
Vehicles: 14505-1 (Principles), 14505-2 (Teq), 14505-3 (human subjects);
14415 (disabled, aged...); 15265 (risk assessment); 15743 (working practices
in cold); 15742 (Combined environments)

Figure 12.3: ISO standard methods for different aspects of climate parameters (based on Figure 1 from Parsons, 2006; p. 370).

Image by C. Berlin.

Figure 12.4: Equipment for measuring WBGT being used in the field by a U.S. Navy seaman.

U.S. Navy photo by Gary Nichols / Released. The image is in the public domain via www.navy.mil.

Climate parameters (objective)

A known tool for measuring climate parameters in non-comfort climates is the "Wet Bulb Globe Thermometer", which measures the "Wet Bulb Globe Temperature Index" (WBGT) as defined by the Heat Stress standard ISO 7243 (Parsons, 2006). The WBGT is measured with:

- A sensor that measures the air temperature (thermometer), shielded from radiation.
- A cylindrical natural wet bulb sensor that measures the temperature under the influence of humidity, covered by a sleeve of wet cotton material.
- A standardized thin, matte, 150mm-wide black globe at whose centre we measure the globe temperature, which measures the impact of radiation on temperature.

These different components (Figure 12.4) allow us to take consideration of wind chill and radiation factors, as well as the air temperature. Different coefficients are used to combine the measurements into the WBGT, depending on whether the measurement is taken indoors or outdoors (since outdoor environments have different radiation and wind chill factors to take consideration of).

Indoors: $WBGT = 0.7t_{nw} + 0.3t_g$

Outside buildings (with solar load): $WBGT = 0.7t_{nw} + 0.2t_g + 0.1t_a$

Where:
t_{nw}: natural wet bulb temperature
t_g: globe temperature
t_a: air temperature

The measured WBGT is compared to a reference value determined by the ISO standard as suitable for the studied activity. The reference value is set so that there is no risk that the body temperature is altered as a result of the activity being performed in that temperature. The standard also states limit values for different durations of activity (Parsons, 2006).

Thermal comfort (subjective)

The standard ISO 7730:2005 describes a way to assess how many people in a population will be satisfied with the temperature at a workplace. The standard applies to healthy men and women in moderate indoor thermal climates, so adjustments must be made for sick people as they may perceive the thermal climate differently. This is a two-step method that takes consideration of a) activity and thermal load (the Predicted Mean Vote, PMV), and b) the proportion of a group of people who will find the temperature too warm or too cold (the Predicted Percentage of Dissatisfied, PPD).

The PPD is calculated as a function of the PMV, so these calculations go together. According to the standard ISO 7730:2005, "The PMV is an index that predicts the mean value of the votes of a large group of persons on the seven-point thermal sensation scale" (ISO 7730:2005 p. 2) which is shown in Figure 12.5.

The PMV is based on studies of people being exposed to different temperatures and then assessing their opinion of the climate on the scale in Figure 12.5. The calculation of PMV is a rather complicated one that takes "metabolic rate, clothing insulation, air temperature, radiant temperature, air velocity

+ 3	Hot
+ 2	Warm
+ 1	Slightly warm
0	Neutral
− 1	Slightly cool
−2	Cool
− 3	Cold

Figure 12.5: The PMV index Thermal Sensation Scale (from ISO 7730:2005).

and air humidity" into account (the detailed equation is in ISO 7730:2005 p.3). In its simplest form, the equation for PMV looks like this:

$$PMV = (0.303* \ e^{-0.036M} + 0.028)L$$

Where M = Activity and L = Thermal load.

Once the PMV is obtained, the PPD can be calculated to estimate the expected number of people who will find the temperature to warm or too cold (predicted percentage of dissatisfied).

$$PPD = 100 - 95*e^{[-(0.03353*PMV4+ 0.2179* \ PMV2)]}$$

Since it is unlikely (due to variations in personal preference) that a majority of the population will be satisfied with the climate (corresponding to a PMV value of 0), the PPD chart frequently shows a U-shaped curve as shown in Figure 12.6.

A rule of thumb is that suitable indoor climate is achieved when -0, 5 < PMV < 0, 5 and the PPD is less than 10%.

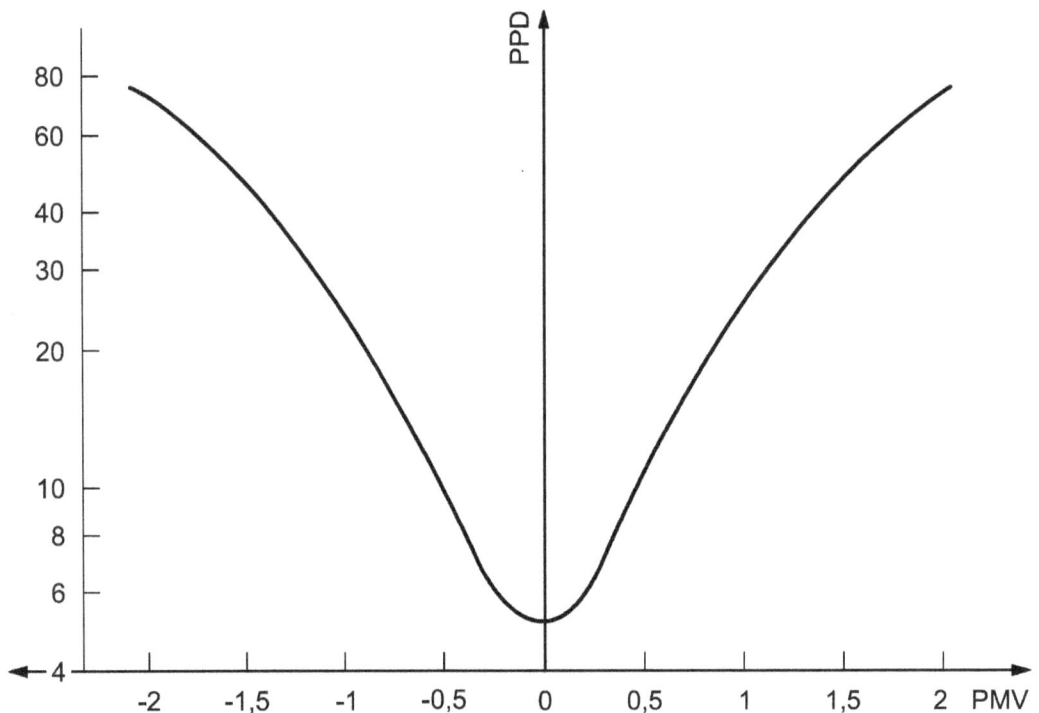

Figure 12.6: Usual appearance of a PPD chart as a function of PMV (ISO 7730:2005).

12.8. Clothing

Changing clothing changes the human's ability to exchange heat with the environment. The insulating properties of different types and amounts of clothing has a definite effect on our ability to work in different climates, due to their ability to limit heat exchange due to radiation and convection, and their interference with evaporation. The insulating capacity is measured in terms of *clo*, a unit of insulation defined in the standard ISO 7730:2005 (Figure 12.6). It is close in magnitude to the R value, an measure used to describe the insulation of housing (thermal resistance), for a particular material or

Garment	I_{clu}	
	clo	$m^2 \cdot K/W$
Underwear		
Panties	0,03	0,005
Underpants with long legs	0,10	0,016
Singlet	0,04	0,006
T-shirt	0,09	0,014
Shirt with long sleeves	0,12	0,019
Panties and bra	0,03	0,005
Shirts/Blouses		
Short sleeves	0,15	0,023
Light-weight, long sleeves	0,20	0,031
Normal, long sleeves	0,25	0,039
Flannel shirt, long sleeves	0,30	0,047
Light-weight blouse, long sleeves	0,15	0,023
Trousers		
Shorts	0,06	0,009
(...)		
High-insulative, fibre-pelt		
Boiler suit	0,90	0,140
Trousers	0,35	0,054
Jacket	0,40	0,062
Vest	0,20	0,031
Outdoor clothing		
Coat	0,60	0,093
Down jacket	0,55	0,085
Parka	0,70	0,109
Fibre-pelt overalls	0,55	0,085
Sundries		
Socks	0,02	0,003
Thick, ankle socks	0,05	0,008

Figure 12.7: Selected values of thermal insulation (in clo), excerpted from table C.2 in the standard ISO 7730:2005.

combination of materials. 1 clo corresponds to 0,155 m² K/W (pronounced "metres squared Kelvin per Watt") and varies between 0 to 3 clo. For reference, a person dressed in 0 clo is naked, while a 3-clo outfit is suitable for someone who is going skiing.

The amount of insulation that allows a human (at rest, and in an environment at 21°C room temperature and 0.1 m/s air velocity) to remain in thermal equilibrium (i.e. no sweating, freezing or change in body temperature) is 1 clo, thanks to the combination of the clothing insulation and the body's heat exchange processes.

12.9. Lighting

Sight and the visible spectrum

Sight is one of the most dominant senses, controlling around 90% of our daily activities (Kroemer, 1997). Our eyes are constantly supplying us with information through visible light. So to ensure the correct information is supplied, lighting is a key consideration in the design of workplaces. Essentially, light is a form of electromagnetic radiation, and the visible spectrum is a portion of the electromagnetic spectrum that can be identified by the human eye, due to its wavelength (Figure 12.8). A typical human eye will respond to wavelengths from about 390 to 700 nm (Starr, 2005).

The process of interpreting an environment through visible light by the brain is called visual perception, and the outcome of this process is known as sight or vision. How we see an object is a combination of light, the object being perceived, the eye and perception. Rays of light from an object or the environment pass through the pupil of the eye and meet at the retina. The light energy is then converted to bioelectric energy, which stimulates the optic nerve to the brain. After a series of impulses and various filtering processes, all the signals are integrated into a representation of the external environment in the brain's cerebral cortex.

While an in-depth understanding of the complex interaction between the eye and the brain during visual perception is outside the scope of this book, an understanding of the different characteristics of light and how to measure them supports the design of healthy workplaces. Studies have shown the impact lighting can have on workplace productivity, with reductions in rejected products and accidents. One such study in an American factory demonstrated a 5% production increase when illumination was increased, combining this productivity increase with the reduced amount of waste resulting from enhanced lighting, cost savings of 24% were achieved (Kroemer, 1997).

12.10. Photometry

This discipline, which deals with the measurement of visible light as perceived by human eyes, is known as photometry (Bass, 1995). There are many different lighting characteristics within the field of photometry, but in terms of ergonomics and workplace design, the three main lighting measurements of interest are:

• Luminous intensity
• Illuminance
• Luminance

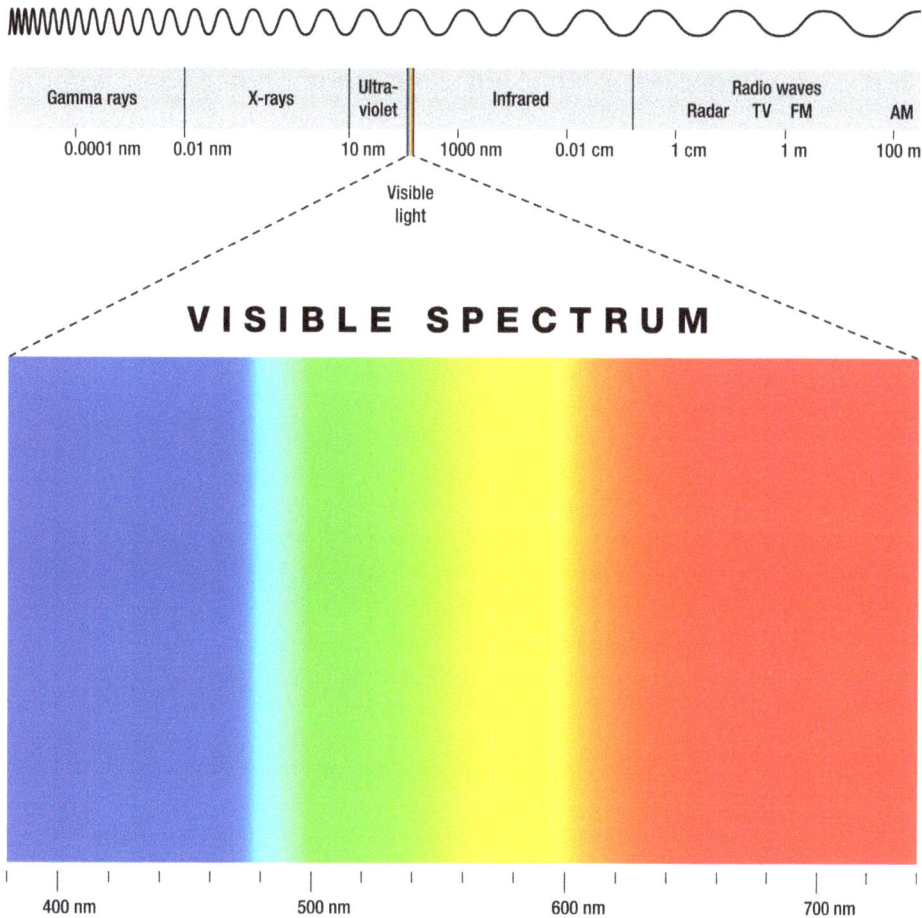

Figure 12.8: Visible spectrum.

Luminous intensity

Luminous intensity is the quantity of visible light that is emitted in unit time per unit solid angle. Historically luminous intensity was measured in terms of the visible radiation emitted by a candle flame, which lead to the name of its measurement unit Candela (cd). Nowadays, 1 cd is defined as 1/683 W/sr at the frequency of 540×10^{12} Hz (Bohgard, 2009).

Illuminance

Illuminance quantifies how well a surface is lit; the light could come from either the sun or artificial light sources. Illuminance is measured in lux and can vary considerably as seen in Table 12.4.

Table 12.4: Outdoor illumination values.

Condition	Illumination (lux)
Sunlight	100,000
Full Daylight	10,000
Overcast Day	1000
Very Dark Day	100
Twilight	1
Full Moon	.1
Quarter Moon	.01
Starlight	.001
Overcast Night	.0001

Luminance

Luminance is the amount of light reflected or emitted from a surface, essentially a measurement of the light power that reaches the human eye, measured in candela per m² (Cd/m²) (Kroemer, 1995; Bohgard, 2009). Excessive luminance in the workplace should be avoided.

12.11. Measuring light parameters

To ensure lighting in the work environment meets standards and recommended guidelines various measuring instruments can be used. Both lux meters and luminance meters use a photo detector to measure illuminance and luminance.

Additional light parameters

In addition to the three photometric quantities we have already introduced, it is also important to consider glare, reflections, contrast and viewing distance when designing workplaces.

Glare

Glare is a visual sensation in which excessive overexposed light impairs vision. In some cases it can impair vision completely, while in other cases it is just deemed uncomfortable and irritating. Glare can occur in four different ways:

- **Direct Glare:** When the light source is so bright the eye can't adjust, so vision is inhibited.
- **Indirect Glare:** When the light source is reflected on shiny surfaces, inhibiting vision.
- **Contrast Glare:** When a significant difference in luminance levels inhibits vision.

• **Adaptation Glare:** When vision is inhibited due to sudden changes between light and dark environments.

Reflections

Reflectance concerns the ability of a surface to reflect light, and is expressed as a percentage (%) of reflected light compared to incident light:

$$\text{Reflectance (\%)} = \frac{\text{Luminance}}{\text{Illuminance}} \times 100$$

Different materials absorb and reflect different amounts of light. Generally in the work environment the walls and ceilings are lightly coloured to enable light to reflect around the room. Equipment and machinery on the other hand, tends to be darker in colour with a matte finish to limit disruptive light reflections.

Contrast

$$\text{Contrast} = \frac{\text{Luminance of object} - \text{Luminance of background}}{\text{Luminance of background}}$$

Contrast is the difference in luminance between two surfaces; this relationship can be described as the brightness of an object relative to its background.

Having significantly different light levels between work areas can be problematic. Colour contrasts between objects and the background can also affect the workplace. Equipment should be coloured differently to the background to ease visual strain for workers. Painting stationary and moving parts in contrasting colours also improves visibility and reduces injury risk (CCOHS, 2013). Colour blindness affects a number of people and can inhibit them from differentiating between certain colours, so special consideration should be taken when selecting the colour and position of safety information and signs.

Visual field

The visual field is the area that can be viewed by both eyes while the head is stationary (Figure 12.9). This area can be subdivided into three sections; the inner field, the middle field and the outer field. The inner field is the zone of sharpest vision, while items within the middle field aren't very clear – they are detected if strong contrasts or movement is used. Usually, little is noticed in the outer field of vision unless movement occurs (Kroemer, 1997).

Viewing distance

When determining the workplace layout and positioning of light fixtures, it is important to ensure that tasks involving detailed viewing are well lit and located in the centre of the workers field of vision, thus enabling good working postures to be adopted while carrying out work tasks.

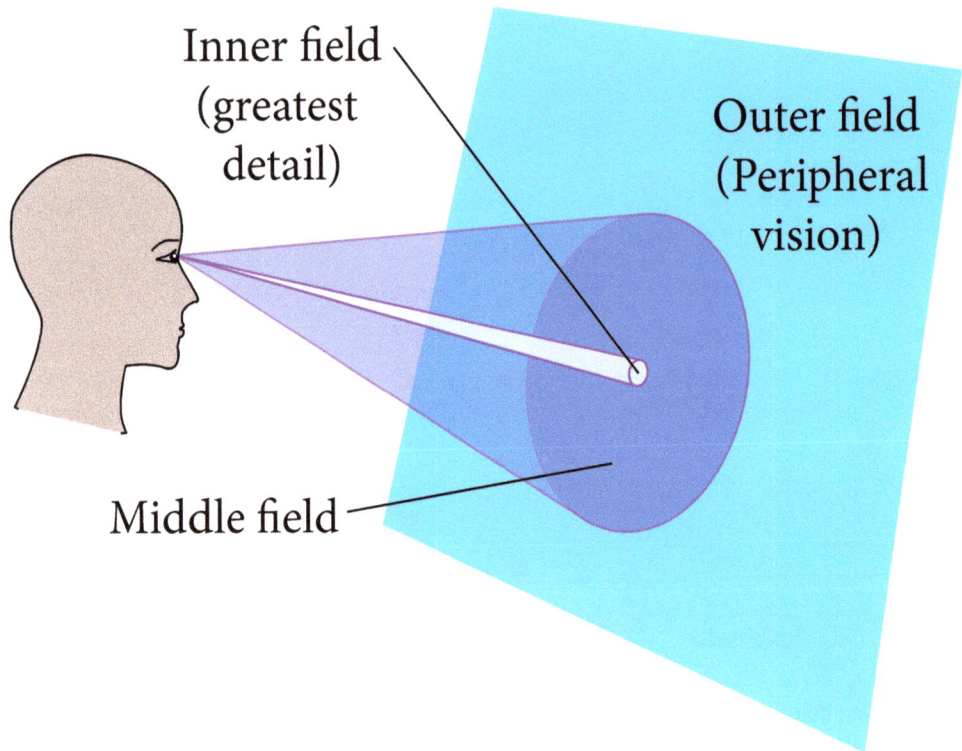

Figure 12.9: Schematics of visual field (adapted from Kroemer, 199 p.297).
Illustration by C. Berlin, based on Kroemer (1997).

12.12. Lighting regulations

A number of standards and regulations have been issued by the Swedish Work Environment Authority concerning lighting conditions in the work environment:

AFS2000:42	Workplace design
AFS1998:5	Working with a computer screen
AFS1993:10	Machine interfaces
AFS1997:11	Cautionary marking
AFS2006:4	Use of work equipment
AFS1998:1	Physical loading, ergonomics
AFS 1980:14	Mental and social work environment aspects
AFS ---	Restaurants, maintenance work, automotive industry, medical controls, explosives, etc.

Glossary

LUMINOUS INTENSITY	Luminous intensity is the quantity of visible light that is emitted in unit time per unit solid angle, Candela (cd).
ILLUMINANCE	Amount of light falling on a surface, measured in Lux (lx).
LUMINANCE	The amount of light reflected or emitted from a surface, measured in candela per m^2 (Cd/m^2).
REFLECTANCE	This concerns the ability of a surface to reflect light, and is expressed as a percentage (%) of reflected light compared to incident light.
GLARE	Excessive or uncontrolled light that impairs vision

12.13. Sound and noise

Sound that can be detected by the human ear is essentially a series of vibrations transmitted through a medium, with frequencies in the approximate range of 20 to 20,000 hertz (Oxford Dictionaries, n.d). Two properties of sound are essential to analysing and designing signals and sound environments; on one hand, *loudness,* measured in deciBels (dB) is the pressure with which the eardrums are physically impacted. The unprotected human ear has physical limitations regarding how much sound pressure it can withstand without sustaining injury, and therefore, loudness is regulated in a healthy work environment. The other property is sound *frequency*, measured in Hertz (Hz), also known as *pitch* or *tone.* Most human ears are differently sensitized to different frequencies; this means that for some pitches, we may require a louder signal to even perceive that a sound is present. For regular human speech, most sounds fall into the frequency range of 250–6000 Hz, where vowels like, a, e and o tend to be low-frequency, and certain consonants (like f, s and th) are high-frequency.

When it comes to aural inputs (what we register through our auditory sensors in the ears), we differentiate sound from noise, where sound is desirable, tolerable signals carrying meaning, whereas noise is unwanted aural input that distracts and causes division of mental resources or discomfort. Noise is subjective and a matter of perception; what one person may consider to be music, another may perceive as noise. There is also the aspect of ambient sound or background sound (such as birdsong, humming machines, murmuring, rolling waves) that our brain is able to filter out or differentiate from meaningful signals. Ambient sound may be borderline to distracting noise, depending on the knowledge and preferences of the individual. In a new environment, initially distracting noise may change over time to ambient sound as we learn to distinguish constantly on-going noises from signals (for example, some people are more skilled than others at filtering out the sound of other people talking in the background, which is often an issue in office landscapes).

Sound signals can be designed to be easily detected (by being different from the ambient sound context), identified and localized. Thanks to binaural (stereo) hearing, we are often able to locate which direction a sound is coming from, since our brain automatically interprets the difference in sound intensity between the two ears (see more in Chapter 5). Sound pressure levels are a logarithmic measure, measured in decibels (dB). Sound signals are commonly used in production environments to alert workers of dangers, e.g. in the form of alarms or sirens.

12.14. Effects of noise

Noise can either be external, coming from outside the building from traffic and other buildings, or internal caused by machines, fans, engines, telephones and people talking. In industrial contexts internal noise can vary considerably depending on the sequence of work tasks. The intermittent clanging of metal, hammering, or hissing spray of paint is commonplace in a production environment. Noise is considered a health hazard as it can trigger hearing loss, cause distraction, mask information, prompt miscommunication and in some cases stress. Prolonged exposure to intense sound can lead to *noise induced hearing loss* (NIHL) over time. For some people, this could occur within a matter of months, while for others the true impact may only be realized years after. Typically loss of hearing is progressive and can also come about through the natural aging process. However, one-off very loud noises can also prompt sudden hearing loss by damaging cells within the ear, known as *acoustic trauma*. A common issue associated with noise is the masking of important information, such as alarms and instructions. Communicating with co-workers can become very difficult in noisy environments, especially when trying to communicate new or complicated technical information. This can be frustrating for workers, and at worst miscommunicated information could prompt severe accidents. Noise can also be very distracting, as you have probably encountered while trying to work in the library and people talking loudly on the other side of the room inhibit you from focusing on your work. Exposure to noise can also have a physiological impact on workers prompting increased stress levels. In addition to the decibel level, the distance of the ear from the sound source and the length of time it is exposed to the sound source are equally important.

12.15. Measuring sound

While it is possible to objectively measure sound, this doesn't always provide information on the disturbing effects of noise in the work environment. So observations and interviews also play an

Table 12.5: Concepts related to sound and noise.

NOISE	• "Unwanted sound" • Health hazard; may contribute to hearing loss • Safety hazard; masks signals, causes irritation and stress
LOUDNESS	• A measure of sound pressure level, measured in Decibels (dB)
FREQUENCY	• The sound property that determines pitch, measured in Hertz (Hz) • The human hearing range is 20 to 20,000 Hz; frequencies below this range are experienced as vibrations • Higher frequencies are the first to be affected by hearing loss • Sounds at extreme pitches (both high and low) cause adverse effects such as pain or nausea
IMPULSE NOISE	• Single short bursts of noise, last less than a second with a peak level 15 dB higher than background noise (Stark, 2003)
AMBIENT NOISE	• The background sound pressure at a given location

important role in determining the levels of uncomfortable noise in the workplace. Sound level meters used to measure sound comprise of a microphone, amplifier, filter and display. A wide variety of meters are commercially available; however, these are of varying quality and accuracy and don't all adhere to international standards, so care should be taken when selecting a tool and where possible only tools with calibration certificates that adhere to international standards should be used.

12.16. Hearing protection

A number of different protective measures exist to reduce the hazardous risks associated with loud noise:

- Hearing Protection Devices (HPDs)
- sound showers (directional speakers)
- sound insulation
- legislation and standards

HPDs are the most commonly used method to protect against high sounds. By wearing earplugs or earmuffs, noise can be damped and the ear protected. However, when worn for prolonged periods of time, these devices can become uncomfortable due to the pressure they exert on contact. *Sound showers* are directional speakers that create a highly focused and directional audio output that can only be heard at a very specific point, so other people in the nearby vicinity cannot hear it. Insulation and sound absorbent materials are often used when soundproofing to convert the energy into another form and prevent it from being reflected around the room. In addition to this equipment, a number of standards and guidelines concerning have been published concerning sound in the work environment, which all companies must adhere to, to minimise potential health risks.

12.17. Vibrations

Vibrations affect our ability to work in both the physical and the mental sense. In a working environment where there is vibration, there is usually also long-term ambient noise which may impair concentration or hearing of important information or signals. In the physical sense, vibrations are a risk because the body tissues and organs absorb the energy from them. Particularly the muscles compensate for the small forces that vibrations expose the body to, both by voluntary and involuntary contractions. If the body is exposed to vibration for a long time duration, this results in excessive low-level static loading, which not only tires the muscles, but also poses a risk to the joints.

As explained in Chapter 4, the joints' contact surfaces are covered with cartilage, to cushion and smoothen the gliding of the bones against each other. Vibrations over long time durations can wear down the layer of cartilage prematurely, causing joint pain and problems. Furthermore, because the cartilage is thinnest at the outer edges, we have the least amount of natural cushioning at the extreme ends of the motion range. This implies that work in extreme postures in a vibrating environment is a particularly hazardous ergonomics risk.

12.18. Whole Body Vibrations

Vibrations appear in many immersive working environments, quite frequently in vehicles such as trucks, buses, ships and forestry equipment, where the body is standing or sitting on a vibrating base. Aside from the risk of injury to muscles and joints, an additional risk factor is that different body tissues have different resonance frequencies, meaning that there is a range of vibrations at which some body tissue will experience local discomfort (Figure 12.10). These resonance frequencies will

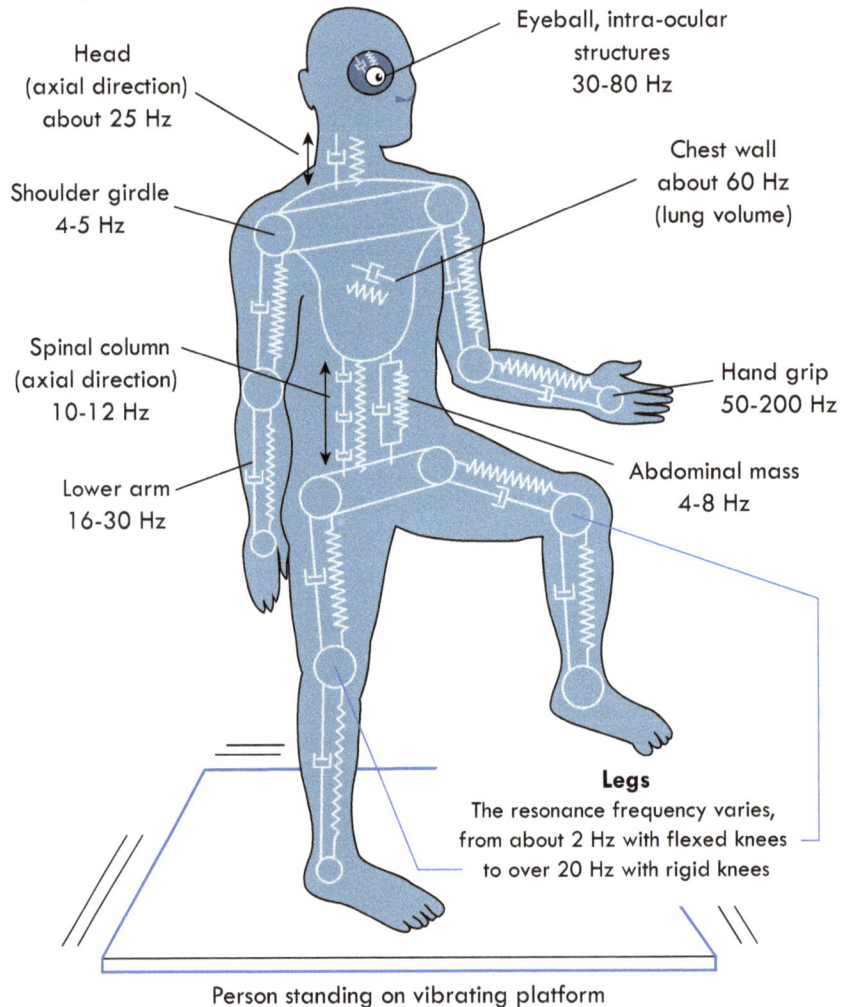

Figure 12.10: Resonance frequencies for different body segments, represented by a simplified mechanical model of a human standing on a vibrating platform.

Image by C. Berlin, based on Rasmussen (1982).

vary from individual to individual but tend to lie within a certain range (for example, the resonance frequency of the shoulder complex typically lies between 4 to 6 Hz, according to Bohgard, 2009 and Rasmussen, 1982).

Not only the joints and muscles are at risk; even the internal organs, eyes, brain and spine are sensitive to vibration at different frequencies. Particularly the eyes at resonance frequency cannot function, leading to impaired visual work due to the vibrating environment.

Low-frequency vertical vibration (lower than 1 Hz) has a particular tendency to cause nausea or drowsiness, depending on the amplitude or intensity and the resonance frequencies of the individual's body segments. This explains why some people can experience motion sickness in a vehicle or ship, while others are lulled to sleep. (Either way, there is a definite impairment to alertness.)

12.19. Hand/arm vibration

In a number of industries, handheld power tools such as chainsaws, pneumatic hammers, grinders and drills are used. As seen in Figure 12.10, frequencies that may be harmful to the arms and hands lie between 16–30 Hz for the lower arm and 50–200 Hz for the hand grip. However, regular use of such tools over a period of time can have severe health consequences and lead to hand-arm vibration syndrome (HAVS), carpal tunnel syndrome (CTS) or arthritis in the wrist or elbow. The symptoms of these syndromes are: vascular, neurological, and musculo-skeletal damage to workers' fingers and hands, tingling and numbness in the fingers, reduced sense of feeling, loss of strength in the hands and episodes of pale, white fingers often triggered by exposure to cold (Work Safe BC, 2013; HSE, 2012). These symptoms can inhibit people carrying out everyday precision tasks such as fastening buttons and pulling zips. Initially these occurrences come and go; however, prolonged continued exposure can make them permanent and irreversible, meaning workers have to change jobs. This damage can be worsened if the workers' hands are cold while they are exposed to vibration.

12.20. Radiation

As mentioned, this book only briefly describes the effects of radiation. Radiation is a mostly invisible environmental factor that has the potential to cause serious long-term ill health effects, and it is important to know something about the range of consequences that may result from radiation exposure. Sources of radiation include equipment, radioactive substances, particles in the air, food, sunlight, lamps, radios and electrically charged materials. Generally, the way to limit radiation exposure is by placing a shield between the source of radiation and the human. The human body absorbs radiation but has the potential to recover from very low doses, as long as sufficient recovery time is allowed between exposures. However, excessive short-term exposure may result in immediate fatal effects. In occupational/industrial hygiene, the remedies to protect workers against radiation are regulation of *time*, *distance* and *shielding* (OSHA, n.d.).

However, it is important to remember that radiation is also very useful; for example, X-rays allow us to non-invasively identify damages in the body; UV radiation can disinfect surfaces and reveal the presence of materials not otherwise visible to the human eye; microwaves allow rapid heating of food materials; and IR cameras can be used to detect motion in places that are too dark for the human eye to see.

Risk assessments are generally made using equipment that can detect electromagnetic radiation (such as radiation detectors or photomultipliers), and measurement units to determine safe levels of exposure are often in terms of absorbed energy per mass unit of body tissue (e.g. the SI unit sievert, Sv, that measures biological effects of ionizing radiation, or SAR, specific absorption rate) or distance to the source of radiation.

Radiation is commonly divided up into ionizing and non-ionizing radiation, where ionizing radiation has the potential to detach electrons from atoms in human tissue, and non-ionizing does not. A general rule of thumb is that non-ionizing radiation sources can be turned off, while this is not possible for ionizing radiation sources.

Ionizing radiation is considered much more damaging to the human body, as physical damage is more apparent at low doses. Short-wave electromagnetic radiation (such as X-rays), charged particles and radiation from radioactive substances fall under this category, and in most working environments it is strongly recommended to eliminate or limit exposure to such radiation sources. Very large exposures to ionizing radiation during a short time period can result in massive tissue damage or cell death, while even intermediate doses may damage cell nuclei and genetic cells to the point where they may grow uncontrollably, resulting in cancer and possibly hereditary damage. A dose larger than 1 Sv received over a short period of time may cause radiation poisoning, an acute condition that can rapidly lead to death.

Non-ionizing radiation, although considered less acutely damaging, can also result in severe ill-health effects, including cancer. Non-ionizing radiation includes electromagnetic radiation with wavelengths corresponding to visible and invisible light (optical radiation) including ultraviolet (UV) and infrared (IR) light; microwaves; and radio-frequency radiation. Optical radiation has physical effects on both the skin and the eyes; in particular, eye damage can result from excessive exposure to UV rays (for example from the sun).

The effects of these two radiation types are summarized in Table 12.6.

Particularly in the nuclear sector, radiation exposure is tightly regulated by international associations such as the ICRP (International Commission on Radiological Protection) and national ones, such as the Swedish Radiation Safety Authority (SSM), both of whom issue design guidelines, limit values and practices to limit exposure to individuals.

Table 12.6: Examples of effects of ionizing and non-ionizing radiation with increasing severity as exposure increases (adapted from Bohgard, 2009 pp. 293–303).

	Ionizing	Non-ionizing
Low to intermediate doses	• Cell damage due to detachment of ions in tissues	• Absorption of energy in human tissues • Altered magnetic fields in the body • Thermal effects (increase in temperature) • Photochemical effects (excited atoms in tissues, resulting in chemical changes)
Higher doses	• Cancer	• Skin pigmentation (UVA rays) • Irritation and potential damage of eyes, snow blindness (UVB rays)
	• Radiation poisoning	• Eye damage (IR radiation) • Cancer (e.g. skin cancer from UV exposure)

Study questions

Warm-up:

Q12.1) Name at least three (each) of the physiological effects of extreme heat and cold.

Q12.2) What are the risks of being exposed to full-body vibration? Name two examples.

Q12.3) Name at least three solutions that counteract noise.

Q12.4) What is glare?

Q12.5) Name three important characteristics of a sound used as an alarm signal.

Q12.6) What is the difference between ionizing and non-ionizing radiation?

Q12.7) Name three disadvantages of protecting humans from extreme temperatures using clothing and protective gear.

Look around you:

Q12.8) Research online for different types of measurement equipment for different work environmental factors (thermal climate, lighting, sound, vibrations and radiation) and try to get a feel for the price ranges that exist – which types of equipment are very costly, and why?

Q12.9) When you encounter a new workplace, try to take note of the conditions of thermal climate, lighting, sound environment, vibration sources and evidence of dust or chemicals. How many of the conditions you observe have an explanation that is connected with the workplace operations? Is the best course of protection to remove the source of exposure, to shield it, or to protect the workers using protective gear?

Connect this knowledge to an improvement project

• If work environment assessments are going to be a frequently occurring activity, it may be a good idea to invest in a toolbox of measurement devices that cover a range of environmental factors. Equipment that measure sound/loudness, temperature and humidity, lighting, vibrations and radiation come in many degrees of sophistication and price ranges, and some are associated with standards that regulate appropriate exposure levels.

• If the workplace you are assessing has an occupational hygienist or work environment representative, they may be a very good informant who can supply good knowledge about that particular work site's conditions, risks and regulations that it abides under. If possible, communicate with that person and discuss solutions that will potentially impact their area of responsibility. This person may also be well informed about chemical and other exposure hazards common to that workplace.

Connection to other topics in this book:

- Many work environmental factors influence the human capability for cognition (Chapter 5) and ability to perform physical work (Chapter 2), both by overloading human senses and locomotive structures and by requiring awkward protection gear and equipment.
- Unfit lighting conditions may cause a worker to adapt their posture to be able to see the work they are performing, which may lead to physical overloading (Chapter 3).
- Psychosocial conditions (Chapter 6) may be affected if workers are in a thermal climate or sensory-overloading environment (with regard to sound and noise) that makes work more strenuous.
- Exposure to vibration may cause long-term serious injuries to different body structures or cause nausea (Chapter 2).

Summary

- Several different factors can cause mental, psychological and physical loading on the body; five main areas are thermal climate, air quality, lighting, sound, vibration and radiation.
- The discipline of occupational hygiene addresses these and many other work environmental factors but includes a wider scope of topics (e.g. chemical exposure, air quality etc.) due to the fact that many occupational hygienists start out as physics or chemical engineers.
- The ideal thermal climate, also known as the comfort climate, is the psychophysical state in which humans experience satisfaction with the climate and so is best for working.
- Suitable clothing for the task and temperature conditions should be worn.
- Suitable lighting that illuminates the working area without causing glare or reflections should be used.
- Sound can be differentiated from noise. Where sound is considered desirable carrying meaning, noise is unwanted, distracting aural input.
- Ambient sound may act as distracting noise, depending on knowledge of the environment and the individual's cognition processes.
- In some workplaces, hearing protection is required to protect workers' ears from noise induced hearing loss (NIHL).
- Vibrations can affect the human's ability to work, both in the physical and mental sense.
- Vibrations are present in a number of work environments, causing injury to muscles, joints and body tissue.
- Low-frequency vibrations can cause nausea and drowsiness.
- In industries using power tools, hand-arm vibrations (HAV) are frequently occurring and can trigger injury, resulting in tingling, discomfort and reduced capacity for tactile feeling in the hand.
- Radiation is predominantly an invisible environmental factor that has potential to cause serious long-term ill health effects; exposure over prolonged periods of time should be monitored.

12.22. References

American Industrial Hygiene Association. (2016). Discover Industrial Hygiene. [Online]. https://www.aiha.org/about-ih/Pages/default.aspx [Accessed: 21 June 2016].

American National Standards Institute, & American Industrial Hygiene Association. (2005). *American National Standard: Occupational Health and Safety Management Systems*. AIHA.

Bass, M. (Ed.), (1995). Handbook of Optics Volume II – Devices, Measurements and Properties, 2nd Ed., McGraw-Hill. ISBN 978-0-07-047974-6

Bohgard, M. (Ed.) (2009). Work and technology on human terms. Stockholm: Prevent. ISBN 978-91-7365-058-8

CCOHS. (2013). Lighting Ergonomics. [Online]. http://www.ccohs.ca/oshanswers/ergonomics/lighting_survey.html [Accessed: 14 Jan 14 2014].

IOHA. (n.d.). What is Occupational Hygiene? [Online] http://ioha.net/faq/ [Accessed: 21 June 21 2016].

ISO. (1982). ISO 7243 ISO 7243, 1982, Hot Environments – Estimation of the heat stress on working man, based on the WBGT-index (wet bulb globe temperature). Geneva: International Standards Organization.

ISO. (2005). ISO/IEC 7730:2005 Ergonomics of the thermal environment – Analytical determination and interpretation of thermal comfort using calculation of the PMV and PPD indices and local thermal comfort criteria. Geneva: International Standards Organisation.

HSE. (2012). Hand-arm vibration at work: A brief guide. [Online]. www.hse.gov.uk/pubns/indg175.htm [Accessed: 14 Jan 2014].

Kroemer, K. H. E. & Grandjean, E. (1997). *Fitting the task to the human: a textbook of occupational ergonomics.* London; Bristol, PA: Taylor & Francis.

Nichols, G. (2010). File: US Navy 100524-N-5328N-671 Cryptologic Technician (Technical) Seaman Antron Johnson-Gray checks the wet bulb globe temperature meter.jpg. [Online] Available from: https://commons.wikimedia.org/wiki/File:US_Navy_100524-N-5328N-671_Cryptologic_Technician_(Technical)_Seaman_Antron_Johnson-Gray_checks_the_wet_bulb_globe_temperature_meter.jpg?uselang=en [Accessed: 28 Dec 2016].

NIOSH. (2013). PtD – Structural Steel Design – Instructor's Manual. [Online] http://www.cdc.gov/niosh/docs/2013-136/pdfs/2013-136.pdf [Accessed: 21 June 2016].

NIOSH. (2015). HIERARCHY OF CONTROLS. [Online] http://www.cdc.gov/niosh/topics/hierarchy/default.html [Accessed: 21 June 2016].

OSHA. (n.d.). INDUSTRIAL HYGIENE Overview [PDF]. Occupational Safety and health Administration Office of Training and Education. [Online]. https://www.osha.gov/dte/library/industrial_hygiene/industrial_hygiene.pdf [Accessed: 21 June 21 2016].

Parsons, K. (2006). Heat stress standard ISO 7243 and its global application. *Industrial Health*, 44(3):368–379. DOI: http://dx.doi.org/10.2486/indhealth.44.368

Rasmussen, G. (1982). Human body vibration exposure and its measurement. Brüel and Kjaer.

Sound. (n.d). In Oxford Dictionaries. Oxford University Press. Oxford University Press. [Online]. http://www.oxforddictionaries.com/definition/english/sound [Accessed: 14 Jan 2014].

Starby, L. (2006). En bok om belysning, Stockholm: Ljuskultur.

Starck, J., Toppila, E. & Pyykko ,I. (2003). Impulse noise and risk criteria. *Noise Health*, 5:63–73

Starr, C. (2005). *Biology: Concepts and Applications.* Thomson Brooks/Cole. ISBN 0-534- 46226-X

Work Safe BC. (2013). Occupational Health and Safety Regulation G.7.14 Vibration Exposure [Online]. Available from: http://www2.worksafebc.com/publications/ohsregulation/guidelinepart7.asp-SectionNumber:G7.14 [Accessed: 14 Jan 2014].

CHAPTER 13

Social Sustainability

Image reproduced with permission from goodluz/Shutterstock.com. All rights reserved.

THIS CHAPTER PROVIDES:

- The place of social sustainability in the context of sustainability as a whole.
- A systems view of how individuals, industries and society relate to each other.
- Design concerns when addressing the well-being, performance and retention of future workforce.

How to cite this book chapter:
Berlin, C and Adams C 2017 *Production Ergonomics: Designing Work Systems to Support Optimal Human Performance.* Pp. 241–258. London: Ubiquity Press. DOI: https://doi.org/10.5334/bbe.m. License: CC-BY 4.0

WHY DO I NEED TO KNOW THIS AS AN ENGINEER?

Sustainability – in any form – is about enabling the present generation of people to live well and fulfil their goals, without threatening the possibility of future generations to do the same. This idea can be scaled up and down, from a lofty global socio-economic perspective all the way down to what goes on in the interaction between a human and a machine. One thing is certain – to understand sustainability, we must understand that its challenges emerge primarily from an interlinked, global and systemic view, which sometimes makes it challenging to apply to isolated sub-system concerns. But it is possible!

The common understanding of these developments remains low. Since Western economies have periodically (and recently) experienced economic recession, the ghost of instable economy and high unemployment makes many people think that corporations are at liberty to treat workers as expendable. This, however, cannot remain a viable business model in the long run, because competitiveness, innovation and increased productivity (not to mention a stable society) can only be achieved with a healthy, motivated workforce willing to stay longer and develop.

In light of this, the concept "social sustainability" becomes relevant to relate to. At the time this book is written, there is unfortunately little agreement in research about what the meaning and scope of social sustainability should be, leaving many companies and practitioners with the responsibility of defining what it means for themselves. For these reasons, the purpose of this chapter is to make the case for ergonomics and human factors as a means to achieve social (and economical) sustainability in production. With good knowledge of human needs and capabilities, engineers can address upcoming social challenges at the local as well as the macro level.

In the future it will remain important for an engineer to be able to persuasively argue for both short-term improvements and long-term supporting of human performance. Without social sustainability awareness, it will become difficult to communicate to other stakeholders that the best workplaces for human beings are part of a sustainable development. To achieve that, you must become aware of how your design skills can align with helping the productivity, quality and efficiency of experienced aging workers high, while at the same time attracting a new generation of workers who will see a positive and attractive future in the places that you design for them.

Furthermore, when you graduate, your ability as a prospective job applicant to identify socially sustainable companies may not only leverage your career – it may also inspire more companies to start taking action towards social sustainability, when they realize that this is the long-term way to attract the best future talent, keep their business operations running long-term and retain a skilled and knowledgeable workforce.

WHICH ROLES BENEFIT FROM THIS KNOWLEDGE?

All of the roles we have mentioned in previous chapters would do well to raise their perspective from the immediate, day-to-day challenges of running production operations, and thinking about what impact they want to have with their work to face the challenges of changing populations, changing understandings about human needs, demands on (and from) workers, and what kind of workplaces should be provided for future generations by those who design them. Understanding that many global challenges interact to form the conditions under which all work is carried out, may seem far away from the craft and method tool box of production ergonomics/human factors at first. But with a systemic top-down view, we can see that planning, leading, investing in, staffing, equipping and continuously improving future workplaces are all ways to ensure that sustainability challenges of the future are met by each work role in their day-to-day activities; when they are all orchestrated towards a common understanding of what makes a socially sustainable workplace, the impact is visible even in the detail-level improvements. Knowledge in ergonomics and human factors can therefore be a valuable asset that helps these roles to ensure that time, money, knowledge and resources are used well to create workplaces that support current workers as well as meet the needs of future generations of workers.

13.1. Upcoming societal challenges

Demographics – who are the future workers?

Demographics – the measurement of populations by number of people – are about to present the working world with a challenging future. Consider the curve shown in Figure 13.1. This chart shows the number of people (sub-divided into age segments) that made up the population of Europe in the year 2012, with numbers including both people being born and migrating into Europe. To the left of the chart are children and young people who have not yet entered the job market, and on the far right are the people nearing retirement.

The majority of those employed in European production industry are between the ages of 35 to 64. As time passes the bars gradually move further and further to the right of the chart as people age (the ones on the far right decrease in height as more elderly people pass away). Typically, the number of people in each age segment does not change much. At the same time, the shape of the silhouette shows the typical recruitment pattern of production industry: the youngest employees are around

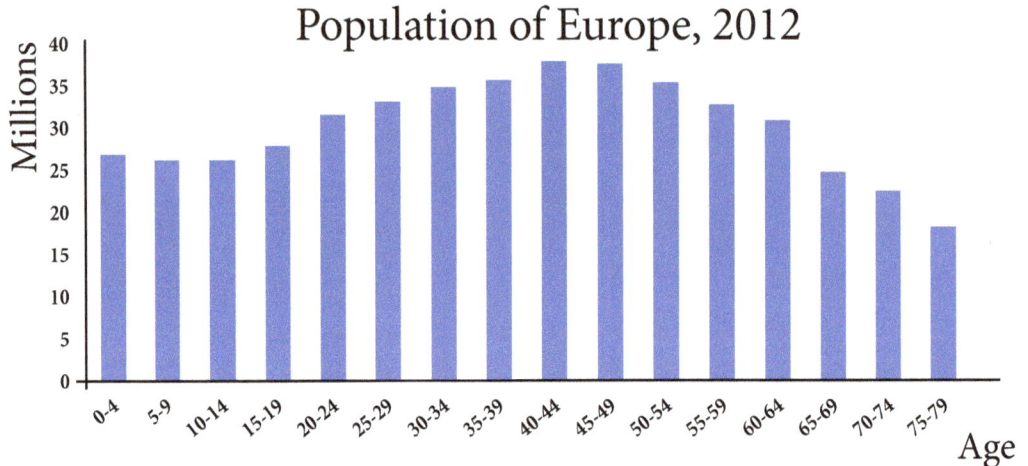

Figure 13.1: The population of Europe in 2012 for different age groups (depicted as bars) according to Eurostat (2012). Typically, the demographics of the industrial workforce has the same silhouette, with the majority being between the ages of 35 and 64.

Image by C. Berlin, based on graph in Berlin et al. (2013).

20 to 25 years old, the majority who stay employed are 35 to 64, and retirement starts around the age of 65 and tapers off to about 74. However, this presents a problem for the future: the sheer numbers of currently employed people aged 40 to 64 are proportionally much more numerous (by tens of millions) than the young people available to replace them, meaning that there is a gap between the number of employed people headed for retirement, and the number of people available to take over their work. Even though this particular chart shows a slight trend change to the left, indicating that more children are being born in Europe, the difference in numbers is still likely to cause recruitment problems, starting in 2020 and continuing into the coming decades. This means that we are experiencing a time in history where the populations in many countries are aging; the proportions of old to young are becoming unbalanced.

At the time this book is written, serious world events in the Middle East have sparked a dramatically increased migration of refugees into Europe, which has been a major transition both at the societal and individual levels; some actors in society choose to frame this as a potential opportunity. If the industrial sector is to continue being a stable source of jobs, productivity and income, these actors argue, integration of newcomers into European society and the labour market is essential. If there is a possibility for them to become employed and trained, their presence will be possible to turn into part of the solution to Europe's demographic challenges. Without functional solutions for labour and educational integration, societal structures may be strained by people who are not able to earn a livelihood due to lack of inclusion, and the lack of staffing in European industries will remain an unsolved challenge. Solutions to these broad challenges must primarily be political and legislative, but workplaces can also prepare for inclusion of new employees – for example with smart communications technology, picture-based instructions, knowledge databases and mentorship between senior and junior employees.

Challenges in countries with aging populations

As populations grow older as a whole, societies face the problem of having decreasing numbers of people working in value-creating jobs that allow export (i.e. production), thereby bringing income and supporting the nation's economy. Another expected development is that the projected number of old people who will require health care will demand that a greater proportion of young people than today work in the eldercare sector. In some countries, an aging population requires that fewer young people will be economically supporting more old people; this may rather soon become the case in e.g. Germany, Italy and China, where the One-child policy of 1979 (Beech, 2013), which previously limited Chinese families to having just one child, was recently phased out (in 2015) as Chinese government officials realized the detrimental and destabilizing effects of having a large aging population and few young people available to enter the labour market.

Figure 13.2 shows the projected changes in age structure between 2010 and 2030 for the German population aged 20 to 64, according to Germany's Ministry of the Interior in 2012.

All in all, this type of demographic imbalance may cause the production industry problems in the future, which can be handled in two ways: the first is to retain older workers until a higher retirement age, and the other is to actively broaden the recruitment pool for new workers. The first approach requires workplaces to be designed to support the needs and limitations of the elderly, so that any age-related impairments (like sight, hearing, and technological unfamiliarity) will have a minimal effect on productivity, and that the know-how of the oldest workers can be put to good use. The sec-

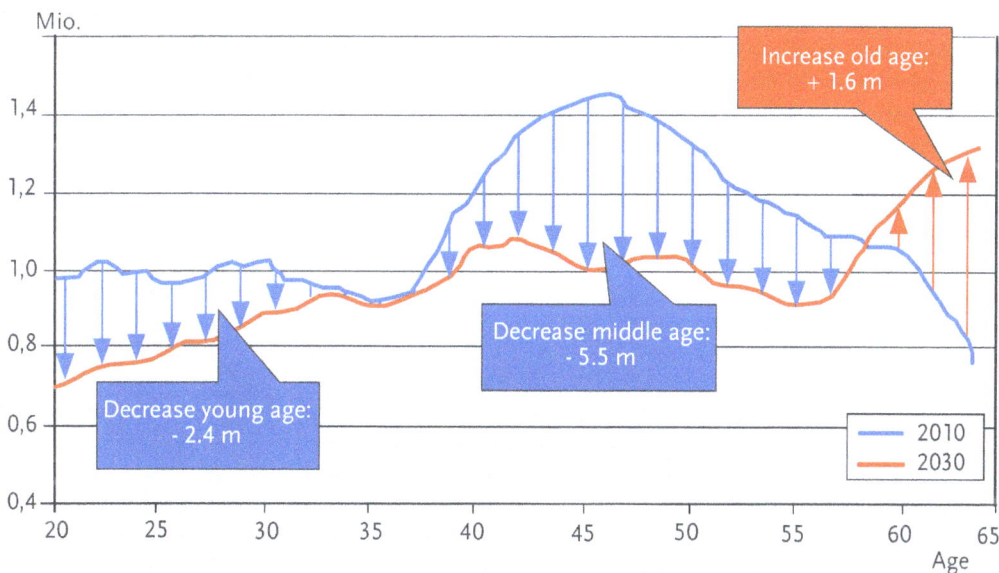

Figure 13.2: Projected differences in the German population's age structure between 2010 and 2030 (Fig. 1 from p. 3 of Richter, Bode and Köper, 2014).

ond approach should not be interpreted merely as attracting a younger workforce; using adaptable design of workplaces, equipment, tasks, training and instructions, it is possible to widen recruitments to a more diverse work pool, including underrepresented genders, "outlier" employees (see Austin and Sonne, 2014) and previously overlooked cultural groups, as well as both older and younger new recruits.

Unemployment

A common counter-argument to the challenges presented above is that Europe has been struggling for a long time with high unemployment rates, particularly for young people. This has been the case for a long enough time to make some people accept this as a universal, unchanging truth: that there will always be more available workforce than jobs. This view tends to foster a cynical outlook that companies are at liberty to treat their workforce badly, as the job market is supposedly full of replacements if they choose to quit, or to move production to low-cost labour countries. Although development suggests that this is changing, it is useful to first define what is meant by unemployment and similar related terms, as this can sometimes cause confusion. Table 13.1 presents unemployment-related terms that may be useful to distinguish between.

These definitions suggest that it is important to differentiate between reasons for not being available to workforce when using the term "unemployed". However, none of the terms in Table 13.1 reflect

Table 13.1: Unemployment-related terms according to ILO (International Labour Office), Eurostat and OECD (combined definitions).

EMPLOYED	People above a specified age who at the time of being surveyed are engaged in paid employment or self-employment (depending on the specific survey, this may sometimes include or exclude part- time employed). Employment (and unemployment) is normally reported for people between 15 to 64 years of age.
UNEMPLOYED	People above a specified age, who at a certain point in time are: • not engaged in paid employment or self-employment • available to start working with short notice, and documented as having been seeking employment (paid or self-employment) during a specified period leading up to the point of being surveyed
LABOUR FORCE	The number of people employed plus unemployed (i.e. who are able and available to work, and may or may not be doing so)
UNEMPLOYMENT RATE	Unemployed people as a percentage of the labour force. (OECD, 2014)
NOT CURRENTLY ACTIVE POPULATION	Persons who are currently not in the labour force, due to any of the following reasons: • attending an educational institution • performing household duties • retiring on pension or capital income • other reasons, including disability or impairment (OECD, 2014)
STUDENT	A person regularly attending an educational institution for systematic instruction at any level of education; not classified as usually economically active (OECD, 2014)

the situation of employers who are unable to find respective employees with a suitable skill set. The mismatch between available workforce and desired skill levels requested by companies has been identified is a major contributor to the massive youth unemployment of the European Union (Mourshed et al., 2014).

Skills gap

Regarding unemployment in Europe, and indeed the rest of the world, the notion that future demographics will result in greater demand for younger workers may be a heartening message, but it needs to be taken with a pinch of salt. What is happening is that the nature of work is also changing, thanks to technology and new knowledge. This means that in spite of the imbalance between young and old, many young people will be unskilled labourers who will continue to be in lesser demand, according to Dobbs et al. (2012).

For production, this means that many jobs are transitioning from thoughtless, menial labour into creative work requiring skill and problem-solving capabilities. The term "skills gap" implies that there is a mismatch between future work roles and the competence of the available workforce today. This gap can be solved in two ways: one is to change the nature of national education at large to prepare young people for work in skilled environments. The other is for individual companies to provide education and training themselves, targeting new employees who do not yet have the required skill set, but are willing to learn and work in that industry. Specific steps towards a better transition from Education to Employment (E2E) have been suggested by McKinsey Center for Government (Mourshed et al., 2014).

Business as usual? Staying competitive in the future

For most industries, from both international and multinational perspectives, what remains a central priority is business and profitability. However, the recognition of the changing times and demographic developments has made it apparent in many countries that future business is dependent on staying innovative and delivering actual high-value products that can be exported. This is becoming evident at policy level in many places in the form of "re- industrialization" initiatives; this denotes the return of manufacturing operations that were previously moved overseas, back to their company's country of origin in order to strengthen economic resilience and ensure jobs and disposable incomes in that country (Foresight, 2013; Westkämper, 2014). In September 2012, the United States Department of Commerce (2012) released the campaign "Make it in America", a manufacturing strategy geared at making more products in America, creating high-skill high-wage jobs and increasing competitiveness, innovation and exports. Similar European goals are being addressed by the Horizon 2020 funding initiative (European Commission, 2014) "Factories of the Future", which supports "manufacturing industry in the development of new and sustainable technologies" in order to keep manufacturing jobs in Europe and addresses many of the aforementioned social sustainability challenges. The drive to be industrially competitive is also visible in the Indian government's campaign "Make in India" (Make in India, 2015), launched in 2014 to convince both multinational and domestic corporations to manufacture products in India, boosting India's domestic job creation and skills enhancement. The movement is also evident in Europe; Swedish production enterprises with international production operations have taken action to bring some production back from foreign countries to Sweden. Two main reasons are cited: the first is that high-level consumer goods sectors benefit from having their production close to product development headquarters, and the second is that production should be close to the customers; for Sweden, the main market tends to be the European Union.

Future needs, wants and expectations

It is important for companies to remember that being an attractive workplace is not a goal in itself, but a means to overcome the challenges of the future, while staying more productive, competitive and profitable.

Returning our attentions to the workforce, it is important to recognize that future needs, wants and expectations are going to be different in the future for different demographic subgroups (such as young, elderly, women, people with families, people at different education levels, etc.). It will become necessary for companies to identify and address those future needs, wants and expectations in order to attract a younger and more diverse workforce, while at the same time supporting the aging population so that the knowledge balance and the company can be retained. At the same time, what is considered a good life and a successful career is a question of societal values, which change over time. For example, according to Accenture (2013), educated women rate work-life balance as a higher priority than salary when seeking a job. Another study by Halkos and Bousinakis (2010) showed that productivity at work decreased when there were stress-related impacts on family life.

People born from the 1980s and later are frequently referred to as "digital natives", due to their life-long exposure to human-computer interfaces and expectations of a high degree of digital connectivity and free access to information. These expectations on technology are coupled to high levels of familiarity with handling interfaces, which perhaps cannot be expected to the same extent from older workers. Furthermore, Karazman et al. (2000) state that a sustained interest in their work is essential to keeping elderly workers.

The business developments of the last couple of decades have displayed behaviours of young professionals "hopping" between careers at different companies, spending just a few years at each position. While this has long been considered beneficial to career development, is hardly beneficial to companies who repeatedly have to re-recruit personnel to replace employees who leave, using up precious resources of time and money and losing valuable experience and competence.

One of the major reasons for creating more "staying power" in a company is to create good conditions for innovation. Innovation is widely cited as the most important way to competitiveness for future industries, but it is dependent on the creativity of humans and of the humans staying at a company long enough to develop skills and knowledge that will make innovative ideas relevant for their specific industrial sector. Although there is a widespread notion that robots and automation still remain a threat to the continued importance of humans in production, more industry leaders are beginning to realize that robots do not have the ability to be innovative.

13.2. Sustainability concepts

Sustainability history

The concept we now know as *sustainability* saw the first light of day in 1987, when the term *Sustainable Development* was famously defined by the United Nations World Commission on Environment and Development (WCED) as:

> "Development that meets the needs of the present without compromising the ability of future generations to meet their own needs."
>
> (WCED, 1987)

This definition came from the commission's report, which is familiarly called "The Brundtland Report" after its chair Gro Harlem Brundtland. At the time, focus was primarily on environmental conservation and responsibly ensuring the continued improvement of economic living standards for developing countries, while conserving the world's natural resources. However, in the years that followed, the wide, abstract concept of sustainable development was elaborated along three tracks:

- Economic: profitability, business growth, meeting market demands
- Environmental: planet, environmental resources, natural heritage
- Social: people, social justice, equity and equal opportunities

The term *Triple Bottom Line* (Elkington, 1998) gained popularity when a popular book on "green business" raised awareness among business managers about how the lifespan of their businesses could be extended if they considered not only the monetary bottom line, but also environmental and social impacts. A catchier, well-known phrase expressing the TBL concept is "People, Planet, Profit". However, the original TBL term was written from the perspective of making progress in "green business" and placed the "people" focus more on social justice.

Glossary of important terms and concepts

SUSTAINABLE DEVELOPMENT	As defined in the Brundtland report (WCED, 1987): "Development that meets the needs of the present without compromising the ability of future generations to meet their own needs."
SOCIAL EQUITY	As defined by the 1996 US President's Council on Sustainable Development (NASA, 1996): "Fair and impartial access to social or public services regardless of economic or social status." Alternative definitions include access to livelihood, education, resources, participation in society and self-determination. May include gender aspects.
SOCIAL EQUALITY	Often considered equal to social equity.
TRIPLE BOTTOM LINE	As defined by Elkington (1998): a business-centred approach to the social, environmental and economic sustainability tracks, the idea is to prepare business on three "accounts": those of "people, planet, profit". TBL has been criticized because of its basic capitalist assumptions and corporate focus, and the fact that social and environmental sustainability cannot be readily translated to the "cash" level of economic sustainability.
SOCIAL RESPONSIBILITY (ISO)	As defined by the standard ISO 26000 (ISO, 2010): how businesses/organizations can operate in the socially responsible way in the society and environment in which they exist, by acting in an ethical and transparent way that contributes to the health and welfare of society.

CORPORATE SOCIAL RESPONSIBILITY (COMMONLY ABBREVIATED CSR)	As defined by Holme and Watts (2000): "a continuing commitment by an organization to behave ethically and contribute to economic development, while also improving the quality of life of its employees (and their families), the local community, and society at large." As defined by the United Nations Environment Programme (UNEP/Setac, 2009): "(…) companies should, at the very least, be held to international standards of human and workers' rights, and that they should consider environmental output regulations when making corporate decisions." Also referred to as "Corporate Social Performance" (Carroll, 1979).
HUMAN CAPITAL	A popular but slightly debated term that conveys the economical view that human beings are an asset in economical systems, partaking in labour and creating value by means of competence, knowledge, cognitive abilities, creativity and personal attributes that make them a valuable human resource. The term is ambiguous because some schools of thought associate it purely with assets in the form of education and cognitive skills, while others do not.
SOCIAL CAPITAL	This concept is centred around social networks between people built on trust, reciprocity and common understandings that allow a society to function through cooperation between groups. Defined by the OECD as "networks together with shared norms, values and understandings that facilitate co-operation within or among groups" (Healy and Côté, 2001).

Social sustainability definitions

From a research perspective, there has not been much convergence in literature on the scope or focus of the term *social sustainability* to date. This makes it difficult to state a universally accepted definition of what social sustainability is, since the angle of which human-related problem to solve often determines the scope. Quite often, different fields of research (including the one of production ergonomics) will decide quite arbitrarily what level of societal inclusion to zoom in on; at present, explicitly stating this range of scope is the best-known way to relate different conceptualizations of social sustainability.

There are also overlaps between social issues and the other pillars of sustainability; for example in ethical sourcing of product materials, housing developments and eradication of world poverty. According to a recent social sustainability literature review by Vallance et al. (2011), some main aspects considered are "inter- and intra-generational equity, the distribution of power and resources, employment, education, the provision of basic infrastructure and services, freedom, justice, access to influential decision making for and general 'capacity building'".

Colantonio (2009) of the *Oxford Institute for Sustainable Development* (OISD) has described some key themes of social sustainability (Table 13.2.). This shows that there is an on-going shift in the understanding of what social sustainability means, moving from human living standards and equity towards work, employment, integration and work-life balance. This reflects the recent focus on demographics-driven changes.

13.3. The ecosystem of social sustainability

In order to achieve lasting solutions, social sustainability initiatives need to be balanced from a number of different perspectives. It is important to realize that the concerns and priorities for different

Table 13.2: Traditional and emerging key themes of social sustainability (Colantonio, 2009)

TRADITIONAL	EMERGING
• Basic needs, including housing • Education and skills • Equity • Employment • Human rights • Poverty • Social justice	• Demographic change (aging and international migration) • Empowerment, participation and access • Identity, sense of place and culture • Health and safety • Social mixing and cohesion • Social capital • Well-being, happiness and quality of life

stakeholders (individual prospective workers, current workers, their families, the company, the community surrounding it and the country as a whole) are tightly interconnected. A change for one stakeholder will affect all the others in terms of expectations, needs, motivations, behaviours (both collective and individual), regulations, infrastructure and values.

Figure 13.3 shows some examples of how the priorities of the individual, industry and society overlap and need to be balanced out in order to secure the right kind and amount of workforce for future

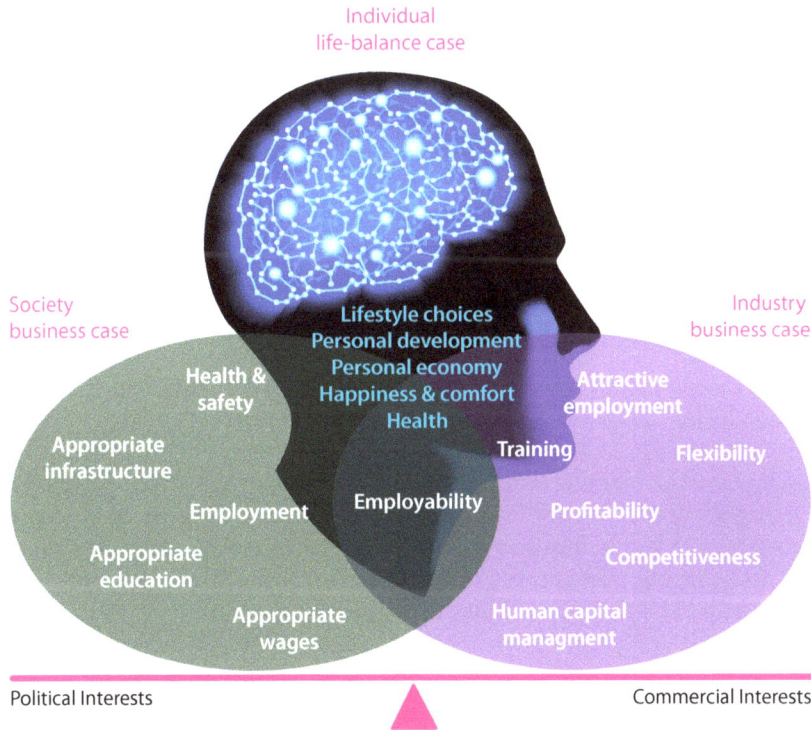

Figure 13.3: The ecosystem of social sustainability, which needs balanced solutions for the needs of the individual, industry and society (SO SMART, 2014).

Image by Vladgrain/Shutterstock.com, with modifications by Elisabetta de Bertti. All rights reserved.

Table 13.3: Examples of social sustainability concerns at different levels.

Individual	• Being adequately paid to maintain a satisfactory lifestyle • Staying healthy, including a sound mental health • Maintaining a healthy work-life balance • Being provided with well-fitted, healthy and understandable workplaces and tasks • Collaborating in a fruitful way with co-workers and management • Developing skills and prominence in one's chosen vocation • Learning and getting support for one's work tasks • Having a purpose in one's social reference system • Developing one's skills and career in one or more companies • Having a safe position to care for family • Having access to living quarters and ways to commute to work • Making conscious choices about pay level and personal motivation (e.g. prioritizing high pay or other motivational factors)
Industry	• Being able to make sound business cases for workplace or organizational change • Staying competitive in a global perspective by managing human capital and talent • Retaining trained and skilled staff • Demonstrating the economic benefits of reducing work-related discomfort, ill health and absence both in direct and indirect costs • Supporting the ability of entry-level staff to learn and perform tasks to a high level of quality • Designing and building workplaces that ensure maximum safety, health and engagement • Creating an exciting work environment and community at the workplace
Society	• Offering and maintaining a steady level of employment for citizens • Encouraging mobility of employable workforce (within and across countries) • Providing adequate general education and training to support industries' need for employable entry-level staff • Decreasing the level of work-related injuries and ill health, avoiding costly economic damage • Providing an attractive infrastructure for business

production. This balance is considered very important from a policy level, and this is highlighted by current initiatives by the European commission (European Commission, 2014.).

Table 13.3 shows examples of some concerns that may affect decisions at individual, company and society levels. It is also important to remember that the decisions at each level are made with respect to different time perspectives. For society, initiatives often need to be made with long-term stability in mind, in contrast with quarterly or yearly goals for a company. For the individual, the timeframe of priorities and decisions will vary, between the short and long-term, depending on personal changes in life (such as starting a family, getting injured, relocating, etc.).

13.4. Social sustainability for work and workplace design

For our purposes of production ergonomics and work design, it can be said that social sustainability concerns the ability of present generations to earn a living and be part of a community they contribute to economic growth (productivity and societal growth) while enjoying well-being and having fair opportunities for remuneration, education, personal development and work-life balance. In this way, the focus is placed on the overlap between social and economical issues. This is an important active choice in order to focus social sustainability action on the future challenges for production companies that we described in the first section of this chapter. To phrase this in terms of our design problem, the shift of focus goes from "decent jobs" to "attractive jobs", in order to attract the most talented work pool and stay competitive.

From that company perspective, social sustainability will then focus on recruiting, keeping and developing employees, so that their personal development benefits their own well-being and the competence level in the company. The two are interconnected, in the sense that personal development contributes positively to well-being. Companies would do well to communicate their focus on this connection in their outwardly communicated vision, company values, activities and benefits or services offered to employees.

13.5. Design for social sustainability

It may be useful to formulate a social sustainability mission statement for any work or workplace design, to guide more detailed design decisions. The authors of this book propose the following mission statement:

Design work and workplaces to achieve the following:

1) Attract and stimulate individuals.
2) Integrate and support groups.
3) Retain and give recognition to teams.

The role of engineers in social sustainability

As engineers, you will find that you are frequently trusted with the mandate to improve and make changes to a workplace. If you are an engineer who is aware of social sustainability and the many benefits that can be reaped from actively designing towards that goal, you have many ways to communicate that awareness, both in what you say and in what you do:

- Carry out work task improvement in a structured manner (analysis and solution building).
- Know which physical, physiological and psychosocial risks to remove from work systems, both the "slow" ones and the sudden ones.
- Know how to provide support for difficult cognitive processes that future operators face.
- Argue for improvements by demonstrating that a human-centred solution can remove many "hidden" costs and productivity barriers.
- Choose suitable equipment from the perspective of a worker who is exposed to loading.
- Know the value of letting workers give input (both for getting ideas and getting acceptance).
- Learn about human needs (see Chapter 6) and think of ways that your engineering skills can fulfil them.
- Speak a language that management understands.

Study questions

Warm-up:

Q13.1) How does the WCED define Sustainable Development?

Q13.2) What is meant by the term CSR?

Q13.3) Name three challenges faced by current European manufacturing companies regarding future staffing.

Q13.4) How is the retention of staff in a company impacted by regularly training them?

Q13.5) In what ways can inclusive workplace design contribute to a workplace being more socially sustainable?

Look around you:

Q13.6) Look at the "Emerging" column in Table 13.2; can you think of ways that current employers might take action to address the concerns in that column?

Q13.7) Can you think of a company that is well-known for wanting to attract a young and talented workforce? List what kinds of incentives they have in place to a) ease work-life balance, b) train and develop their staff and c) support and encourage innovativeness and creativity.

Connect this knowledge to an improvement project

• Identify the "future generations" perspective in your workplace improvement projects and try to make a statement about the projected impacts on current and future workers in the long term. What actions can be implemented to ensure that they are protected and motivated to continue working there?

• Reflect on the impacts on individuals, the work team, the company and society in different improvement projects. How far do the effects of a workplace improvement spread?

• In any project, review the overviews of human needs described in Chapter 6 and see if fulfilling them may give rise to new ideas for improvement potentials.

Connection to other topics in this book:

• A socially sustainable workplace is a very wide concept, so using the knowledge from all of the preceding chapters can contribute to its fulfilment. However, it is important to apply a holistic and systematic approach to avoid optimizing one aspect possibly at the expense of others, so it is important to understand how the different areas of ergonomics overlap and interact with each other.

- The focus on teams, stakeholders and worker involvement in social sustainability discussions make the topics of psychosocial factors and worker involvement (Chapter 6), economic aspects of ergonomics (Chapter 11) and the relation of different stakeholders to the different areas of knowledge (the "roles" section at the beginning of each chapter) particularly relevant.

Summary

- Future demographic developments are likely to have a significant impact on the available workforce for future factories.
- A probable scenario is that there will be a larger proportion of elderly people in many Westernized economies, alongside a shortage of skilled, young workers willing to choose a career in the production sector.
- The persistent problem with high youth unemployment in Europe contrasts with the demographics problem, but can be explained with the *skills gap* concept and the fact that unemployment can be defined in many different ways.
- Future factories must address the challenges of becoming more attractive workplaces, as well as adapting to the needs of an aging workforce.
- Sustainability as a whole is commonly defined as having three tracks: economical, environmental and social sustainability.
- The history of sustainable development starts in the late 1980s and has prior to now had a greater focus on the environmental and economic sides.
- The current demographic developments demand that production companies start to focus more on the overlap between economical and social sustainability.
- Social sustainability issues can be analysed and solved on the individual, industrial and societal levels. Ignoring any of these levels may lead to unbalanced solutions.

13.6. References

Accenture (2013). International Women's Day 2013 Defining Success 2013 Global Research Results. [Online] Available from: http://www.accenture.com/SiteCollectionDocuments/PDF/ Accenture-IWD-2013-Research-Deck-022013.pdf [Accessed 13 Jan 2014].

Austin R. D. & Sonne, T. (2014). The Case for Hiring "Outlier" Employees. Weblog. [Online] Available from: http://blogs.hbr.org/2014/01/the-case-for-hiring-outlier-employees/ [Accessed 13 Jan 2014].

Beech, H. (2 Dec 2013). Why China Needs More Children. *TIME Magazine.* [Online] Available from: http://content.time.com/time/subscriber/article/0,33009,2158110-2,00.html [Accessed 13 Jan 2014].

Berlin, C., Dedering, C., Jónsdóttir, G. R. & Stahre, J. (2013). Social sustainability challenges for European manufacturing industry: attract, recruit and sustain. In *IFIP International Conference on Advances in Production Management Systems* (pp. 78–85). Berlin, Heidelberg: Springer.

Carroll, A. B. (1979). A Three-Dimensional Conceptual Model of Corporate Performance. *The Academy of Management Review*, 4(4):497–505

Colantonio, A. (2009). Social Sustainability: Linking Research to Policy and Practice. [Lecture] Oxford Brookes University, 26–28 May.

Dobbs, R., Madgavkar, A., Barton, D., Labaye, E., Manyika, J., Roxburgh, C., Lund, S. & Madhav, S. (2012). The World at Work: Jobs, Pay and Skills for 3.5 billion people. McKinsey Global Institute [Online] Available from: http://www.mckinsey.com/insights/employment_and_growth/the_world_at_work [Accessed 13 Jan 2014].

Elkington, J. (1998). *Cannibals with Forks: The Triple Bottom Line of 21st Century Business*. Gabriola Island, BC Stony Creek, CT: New Society Publishers. ISBN: 0865713928

European Commission. (2014). Research and Innovation Participant Portal: Call for Factories of the Future. [Online] Available from: http://ec.europa.eu/research/participants/portal4/desktop/en/opportunities/h2020/topics/2183-fof-04-2014.html [Accessed 17 Jan 2014].

Eurostat. (2012). Population on 1 January by five years age groups and sex. [Online] Available from: http://appsso.eurostat.ec.europa.eu/nui/show.do?dataset=demo_pjangroup&lang=en [Accessed 15 Jan 2014].

Foresight. (2013). The Future of Manufacturing: A new era of opportunity and challenge for the UK. Project report. The Government Office for Science, London.

Halkos, G. & Bousinakis, D. (2010). The effect of stress and satisfaction on productivity. *International Journal of Productivity and Performance Management*, 59(5):415–431.

Healy, T. & Côté, S. (2001). *The Well-Being of Nations: The Role of Human and Social Capital. Education and Skills*. Organisation for Economic Cooperation and Development, 2 rue Andre Pascal, F-75775 Paris Cedex 16, France.

Holme, R. & Watts, P. (2000). Corporate Social Responsibility: Making good business sense. World Business Council for Sustainable Development [Online] Available from: http://research.dnv.com/csr/PW_Tools/PWD/1/00/L/1-00-L-2001-01-0/lib2001/WBCSD_Making_Good_Business_Sense.pdf [Accessed 15 Jan 2014].

ISO. (2010). ISO 26000. Social responsibility. [Online] Available from: http://www.iso.org/iso/ home/standards/iso26000.htm [Accessed 15 Jan 2014].

Karazman, R., Kloimüller, I., Geissler, H. & Karazman-Morawetz, I. (2000). Effects of ergonomic and health training on work interest, work ability and health in elderly public urban transport drivers. *International Journal of Industrial Ergonomics*, 25(5):503–511. ISSN 0169-8141

Make In India. (2015). Make In India. [Online] Available from: http://www.makeinindia.com/home [Accessed 20 Nov 2015].

Mourshed, M., Patel, J. & Suder, K. (2014) Education to Employment: Getting Europe's Youth into Work. McKinsey Center for Government. [Online: Report] Available from: http://www.mckinsey.com/Insights/Social_Sector/Converting_education_to_employment_in_ Europe?cid=other-eml-alt-mip-mck-oth-1401 [Accessed 13 Jan 2014].

NASA. (1996). Sustainable Development Indicator (SDI) Group SDI Inventory, Organized by Issue Working Draft, Version 3, October 8, 1996. [Online] Available from: http://www.hq.nasa.gov/iwgsdi/ISS_SDI_Equity.html#Social%20Equity [Accessed 16 Jan 2014].

OECD. (2014). Glossary of statistical terms. [Online] Available from: http://stats.oecd.org/ glossary/ [Accessed 13 Jan 2014].

Richter, G., Bode, S. & Köper, B. (2014). Demographic Changes in the Working World. [Online] Available from: http://www.baua.de/en/Publications/Focus/article30.pdf?__blob=publication File&v=4 [Accessed 15 Sept 2016].

SO SMART. (2014). SO SMART Eco System.[Online] Available from: http://sosmarteu.eu/ [Accessed 20 Nov 2015].

UNEP/Setac. (2009). Guidelines for Social Life Cycle Assessment of Products. United Nations Environment Programme. [Online] Available from: http://www.unep.fr/shared/publications/pdf/ DTIx1164xPA-guidelines_sLCA.pdf [Accessed 17 Jan 2014].

United States Department of Commerce. (2012). Fact sheet: Make it in America Challenge. [Online] Available from: http://www.commerce.gov/news/fact-sheets/2012/09/25/fact-sheet-make-it-america-challenge [Accessed 13 Jan 2014].

Vallance, S., Perkins H. C. & Dixon J. E. (2011). What is social sustainability? A clarification of concepts. *Geoforum,* 42(3), June 2011:342–348. ISSN 0016-7185

WCED. (1987). World Commission on Environment and Development (WCED), Our Common Future. Oxford: Oxford University Press.

Westkämper, Engelbert. (2014). *Towards the Re-industrialization of Europe: A Concept for Manufacturing for 2030,* Springer-Verlag Berlin Heidelberg. ISBN 978-3-642-38502-5.

Bibliography

Docherty, P., Kira, M. & Shani, A. (2009). *Creating Sustainable Work Systems: Developing Social Sustainability.* New York, NY: Routledge.

IISD. (2012). Sustainable Development Timeline. [Online] Available from: www.iisd.org/pdf/2012/ sd_timeline_2012.pdf [Accessed 3 March 2013].

Rosling. (n.d.) Gapminder.org. [Online] Available from: http://www.gapminder.org/ [Accessed 19 Jan 2014].

Notes for Teachers

How to use this book

The philosophy of this book is that the subject of ergonomics and human factors is best taught to engineering students by letting them exercise their skills practically and develop an analytical eye when faced with a real work environment improvement situation – as opposed to learning through rote memorizing and formal exams. Ideally, we encourage teachers to set up the course curriculum to apply the knowledge and topics in this book to a workplace re-design case. This may be a fictive or "real" workplace (meaning tangible and observable; could be in a real industrial workplace or in a lab), where it is possible for students to use data collection and analysis (using the methods in this book) to determine a current state of the workplace, listing improvement potentials. Then, they should use their knowledge of ideal design principles (also in this book) to devise a change project to support human well-being and system performance, and then bring it to a stage of theoretical or practical proof-of-concept that can convince an audience of peers (and the teacher) that the proposal is feasible, from a practical and economic point of view. (This is how we have been teaching the subject, in the form of a seven-week project based on an assembly workstation rigged up in a lab.)

Before this book came into being, its contents had for several years been taught to students in a course at Master of Science level at a Swedish technical university, usually after 2–3 years of production

How to cite this book chapter:
Berlin, C and Adams C 2017 *Production Ergonomics: Designing Work Systems to Support Optimal Human Performance*. Pp. 259–262. London: Ubiquity Press. DOI: https://doi.org/10.5334/bbe.n. License: CC-BY 4.0

engineering studies. The course curriculum was based on all students forming groups of four or five and tackling a workplace improvement project like the one briefly described above. All these aspects allowed the students to exercise much of the book's contents (plus those of another book about work design, the classic tome *Maynard's Industrial Engineering Handbook* (Zandin, 2001); that book covered aspects of productivity measurement, time studies, etc. that are not covered by this book).

This backstory does not exclude a "younger" target audience or a different course setup altogether, but these teachers' notes aim to stimulate the students towards higher levels of cognitive ability (according to Bloom's taxonomy of cognitive abilities; Anderson and Krathwold, 2001), e.g. evaluation, reflection and creation/synthesis, and our experience is that our tried-and-tested improvement project has served this purpose well. At the end of the course, students overall feel that they have grasped the knowledge and are able to act on it independently, thereby making them more capable, analytical and creative as engineers[1]. Another intention is to prepare engineering students to tackle open-ended problems that may be solved in many different ways. This of course requires some self-assurance on the part of the students, as there is no "textbook answer".

On an individual level, the topics can be examined according to the same philosophy as stated above using open-ended essay questions that stimulate students to seek out their own information (e.g. by independently searching for examples, case studies and equipment listings in literature or online) and to use a critical eye to recognize improvement potentials in their own surroundings. This requires the teacher to be observant of the students' analytical approach and process and reflective abilities, rather than simply marking answers as "correct" or "incorrect". This individual research-writing format may be ideal as preparation for discussion seminars. Therefore, the study questions at the end of each topic chapter in this book aim primarily to stimulate students to consider how the knowledge may be applied to a problem scenario.

A further ambition is to have this book serve as a handy reference for students in their future application of ergonomics and human factors in their workplace improvement practice. For this reason we introduce the notion that engineers may take on a number of different roles in their future working life, and that each of those roles may be primarily concerned with analysis and problem solving on many different system levels.

Therefore, it is advisable to alert students to the fact that they are being prepared for a variety of future work-life scenarios, where they themselves may play a variety of roles and will also encounter other roles that may have different priorities and will require different types of evidence to be convinced that an improvement proposal is worthwhile. It is ideal if students are guided towards adopting the perspective of several different stakeholders during the course, in order to understand the main concerns of those stakeholders and be able to communicate with them effectively in the future.

Below, we point out the consistently repeated elements of each chapter, to further clarify their perspective and pedagogical intent:

"Why do I need to know this as an engineer?"	This text aims to appeal to the student's (perhaps fuzzy) idea of what it will be like to work as an engineer in the future, and in particular introduces the scenario of acting as a workplace improvement agent. These short texts discuss how the chapter's knowledge benefits worker performance, productivity and (when applicable) how acting on the knowledge can be a good business case.

The Roles	These "characters" presented in the Introduction re-appear at the beginning of each topic chapter to further emphasize that improving a workplace is usually a team effort, with many different work roles put in charge of overlapping areas of responsibility that can affect workplace ergonomics. The idea is to increase the awareness that each role may seek different types of "key evidence" to be convinced that an ergonomics intervention is worthwhile, will have a desired effect and will target the appropriate concerns.
Study questions	Here, a way to approach and absorb the material individually is aided by a few warm-up questions that guide the student towards seeking the answer in a particular section in the book, followed by a couple of "look around you" questions that aim to train their observational, reflective and analytical abilities. An answer guide is also provided at the end of the book.
"Connect this knowledge to an improvement project"	Similar to the "Design for…" guidelines, these bullet lists provide specific advice that encourages students to combine data collection and analysis (as appropriate) during different stages of an improvement project.
"Connection to other topics in this book"	This section indicates the relation and overlap or interaction between the present chapter and the others in the book; quite often, there are ripple effects of some aspects of ergonomics that end up affecting multiple other aspects. These chapter elements reveal the connections.
Summary	This element serves as a quick reminder of the scope that each chapter covers.
"Design for…"	Compiled in Part 3, these bullet lists offer the student a checklist of "design ideals" that can be useful if an ergonomics intervention is meant to target specific improvement potentials. It is advisable to have students study these and make sure they can explain the underlying theoretical reasons *why* the recommendations are good advice.

Notes

[1] The ideal level of knowledge, skills and competence that we want this book to support corresponds to the European Qualifications Framework (EQF) levels 3 to 6 (European Commission, 2016). However, these levels can only be truly supported through a well-planned curriculum, and should not be expected as a result of merely reading this book.

References

Anderson, L. & Krathwohl, D. (2001). *A Taxonomy for Learning, Teaching, and Assessing: A Revision of Bloom's Taxonomy of Educational Objectives*. New York: Longman.

European Commission. (2016). Descriptors defining levels in the European Qualifications Framework. [Online] Available from: https://ec.europa.eu/ploteus/en/content/descriptors-page [Accessed 20 June 2016].

Zandin, K. B. (2001). *Maynard's Industrial Engineering Handbook, 5th Edition*. [Online] New York: McGraw Hill. Available from: http://accessengineeringlibrary.com/browse/ maynards-industrial-engineering-handbook-fifth-edition/p2000a1fc99706.9001 [Accessed 16 January 2014].

PART 3

Workplace Design Guidelines

Design for the human body

- Minimise tasks where the employee has to work with hands above shoulder level.
- Design variety into work tasks to avoid prolonged periods of static work.
- Provide sufficient space so that operators are not forced to assume bad postures to carry out tasks.
- Minimise highly repetitive motions in extreme positions; where this isn't possible, provide suitable rest and recovery time.
- Encourage employees to pay attention to their posture while carrying out work tasks, keeping their feet planted firmly on the ground with the knees, hips and shoulders in line.
- Avoid frequent or static bending of the neck; especially important for environments where screens/tablets/smart phones are frequently used.
- Factor in breaks with muscle relaxation.
- Avoid working tasks at low heights.
- During lifting tasks the majority of the load weight should be taken by stronger leg muscles rather than the weaker upper torso.
- Precision work is suited to the hands while tasks involving loading are more suited to the legs.
- Avoid tasks involving a high degree of twisting or bending of the spine.
- Combine static and dynamic loading tasks.
- "The next sitting position is the best sitting position" – so design to allow variation in sitting posture.

Design of hand tools

- Natural hand grips and the functional position of the hand should be used when working with hand tools – avoid bending and twisting.
- Consider the level of force and precision needed for the task, and reflect this in the tool shape and weight (the moment of inertia should be close to the wrist for best balance).
- Design for low muscle tension during prolonged work.
- Provide large grip areas with low and equal pressure distribution on the hand and optimized force transfer through handle.
- Avoid sharp edges that may result in discomfort or pinch injuries – think of safety!
- Consider work environmental factors that may affect tool use, such as climate, vibrations, lighting, etc.
- For extreme climates, provide thermally isolated grip.
- Avoid vibrations, particularly in the injury range of 5 to 2000 Hz.
- Consider static loads vs. impulse loads and support the body in handling these.
- Be aware of the needed comfort-working space necessary for correct gripping.
- Design for easy use in narrow spaces if needed.
- Design for stability and pressure areas.
- Ensure low friction.
- Enable use in different positions and with different hand grips.
- Design tools to be easy to control and adjust without changing grip and/or while wearing gloves.
- Design tools to be possible to adjust to different hand and arm sizes, and to enable use with both hands if needed.

- Ensure that tool surfaces provide appropriate friction for a safe grip.
- Design tools to be resistant to chemicals, blood, etc.

Design for anthropometry

- Design for populations rather than individuals – consider the pool of people you would want to be able to work in your workstation.
- Workplaces should be designed for the 5th–95th percentile, enabling the exertion of muscular strength with most efficiency and least effort.
- Consider biological variation between ages, genders and nationalities.
- Make sure that the main objects most frequently used by workers are within close range; placing items in the order they are used can be a good layout option.
- Exclude as few people as possible.
- The "average person" does not exist – instead, determine your "critical user" whose needs are important to meet in your design.
- Whenever possible, design for adjustability.
- Design for extreme individuals when relevant (for example, the height clearance of doors).
- Use working heights 50–100mm below elbow height.

Design for cognitive support

- Support and enhance human senses; provide good lighting, minimize noise, use haptic signals, provide redundancy (overlap) in sensory stimuli.
- Minimise the need for keeping too much information in the short-term memory.
- Aid perception using visual cues, pattern recognition, consistency in design.
- Avoid information overload.
- Use standardized work.
- Provide each workstation with work instructions.
- Use poka yoke methods; pick by light or voice, or andon.
- Simplify product designs to aid assemblers (DFA).

Design for psychosocial health and worker involvement

- Make "creating the right conditions for other people to perform" your overall design mission.
- Minimise the occurrence and effects of negative stress.
- Provide support for workers to handle stress – consider the cognitive needs of novices and experts.
- Consider human needs in a wide perspective.
- Strive to match the levels of control, demands, decision latitude, support and supervision to the individual's skill, experience and maturity to make their own decisions.
- Use design models at different stages of the design process to stimulate the workers to discuss and give ideas – this fosters solution ownership, innovation and acceptance.

Design for materials handling

- Clearly display material in a logical way to provide operators with cognitive support.
- Position material so that bending and twisting of the spine is minimized.
- Minimise the time spent handling material for assemblers (less distance and transportation, less grasping time and less searching).
- Reduce space taken by storage containers at the assembly line.
- Where possible, allow the operator to only do assembly operations, no intermittent operations.
- Design material façades with the most frequently used component in the "sweet spot", where loading, bending and twisting is minimal.

Design for thermal climate

- Change the activity/task.
- Determine suitable clothing.
- Find radiation sources and lower their temperature.
- Change the air humidity.
- Insulate exposed surfaces with high/low temperature.
- Decrease the exposure time.
- Design variation into the work to even out exposure.
- For learning, concentration and mental work, aim for 20–22 °C.
- For creative work, aim for 23–26°C.
- Inform yourself about individual preferences; ask operators and offer various solutions.

Design for good vision

- Avoid significant differences in luminance within the room as these can create sudden contrasts that are difficult to adapt and transition between.
- Use a combination of general lighting and specific task lighting (e.g. spotlights) for individual workstations.
- Light should be directed so that working areas aren't in shadow.
- Where possible, maximise the amount of natural daylight in workplace; however, windows should be fitted with blinds to minimise glare on sunny days.
- Paint walls and ceilings with light colours to allow light to reflect.
- Ensure safety signs utilise suitable colour contrasts, and consider those who are colour blind.
- Create a contrast between objects and the work environment, (e.g. use different colours for doors and other functional furniture).
- The working field should be brightest in the middle and get gradually darker towards the edges. A 5:3:1 luminance ratio is recommended, where the lighting in the inner field of a worker's vision is five times that of the outer field of vision, while the surrounding field is three times that of the outer field of vision (see figure below).

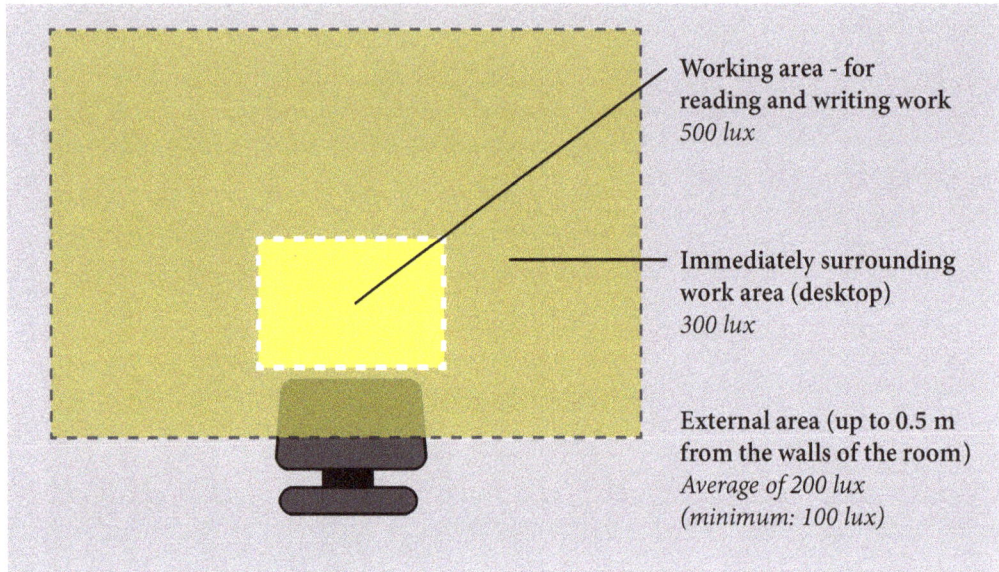

Working area - for reading and writing work
500 lux

Immediately surrounding work area (desktop)
300 lux

External area (up to 0.5 m from the walls of the room)
Average of 200 lux (minimum: 100 lux)

5:3:1 Luminance ratio.

Illustration by C. Berlin, based on Starby (2006).

Design for healthy sound environments

• Select equipment and machinery that emits as little noise as possible.
• Enclose loud sources of noise in soundproof cases.
• Fit walls and ceilings with sound absorbing panels to reduce the amount of reflected noise (echo effect).
• Sound dampening should be done at the source; however, if no other means are possible then hearing protection should be used, but only as a last resort.
• Provide employees with hearing protection if noise exposure exceeds 80 dB.
• Provide employees exposed to noise with frequent hearing checks so any damage can be identified quickly.
• Ensure that emergency alarms are louder and at a different frequency than ambient background noise.

Design to minimize whole-body vibration risk

• Design suspended seating that is adjustable for the worker's weight and has a vibration-damping mechanism.
• Avoid sudden load changes (picking up, dropping off).
• If possible, administer breaks in a non-vibrating environment.
• Fit vibration-damping mechanisms where possible.

- Maintain shock absorbers on vehicles.
- Isolate booths/cabs, etc., by setting them on their own separate foundations.
- Instruct workers not to jump out when exiting equipment or a vehicle, as the intervertebral discs may have been softened by the vibrations, causing greater vulnerability to shock and damage to the body.

Design to minimize hand-arm vibration risk

- Ensure that tools are properly maintained, serviced and adjusted.
- Use balancers and suspended mounts to reduce vibration of the hand.
- Replace anti-vibration mounts and suspended handles before they deteriorate.
- Keep tools such as chisels sharpened.
- Ensure that rotating tools are dynamically balanced.
- Grip the tool handle with the least hand strength possible.
- Cover handles with an insulating layer to offer both thermal and vibration isolation – especially for metal handles.
- Provide anti-vibration gloves to the workers.
- Replace old tools with newer ones with lower vibration.
- Eliminate or reduce the need for vibrating tools, e.g. using automation or changing processes.
- Limit the worker's daily vibration exposure, e.g. by job rotation.

Design for social sustainability

- Design attractive workplaces that address the physical, cognitive and psychosocial needs of employees.
- Support the physical and cognitive limitations of aging workers.
- Design work tasks and workplaces to attract future generations.
- Design so that other cultural or language backgrounds will have minimal impact on productivity and efficiency.
- Design to attract individuals: be informed about what is sought after and appreciated by different ages, genders, nationalities, educational backgrounds and technology proficiencies.
- Design to integrate and solidify groups: strengthen and support team roles, belonging, communication and visualisation of the work.
- Develop the competence of employees by providing suitable training, education and challenges.
- Ensure that work stimulates and engages employees.
- Ensure that ethical standards and international norms are followed.
- Support employees' work-life balance.
- Enhance employees' quality of life.

Answer Guide to Study Questions

Basic Anatomy and Physiology

Q2.1) (Skeletal) muscles, bones, and joints.

Q2.2.) Active: muscles. Passive: the spine (consisting of vertebrae), gelatinous discs, cartilage and ligaments.

Q2.3) During the day, the spinal structures are compressed by the vertical load of our upper body weight (plus external loading), leading to a gradual flattening of the intervertebral discs. Lying down overnight allows the discs to regain their shape, allowing better dampening of forces and shocks.

Q2.4) The shoulder area is a very complicated combination of four different joints and many small muscles, tendons and fascia that are tightly interwoven. Shoulder health is dependent on these structures being well-balanced, so any repetitive or monotonous strain disrupts this balance and causes pain, discomfort and sensitivity to injury.

Q2.5) Our leg muscles are the strongest muscles we have and are developed to exert large forces. The back muscles are weaker and meant to hold up the body rather than to handle external weights; it is also easier to engage the leg muscles in their entirety, which is not the case for coordinating all back muscles.

Q2.6) At the ends of our motion range, the cartilage coverings at the ends of our joints are the thinnest, and the internal pressure is the highest on structures and on passageways with nerves and blood vessels, so any additional external loading on top of an extreme position is a very weak starting point for performing any work.

Q2.7) When training intentionally, the maximal force exertions and exhaustion are complemented at will with rest periods, which allow the body to replenish blood flow and oxygen and relax muscles. In this way, the exercise is not harmful as the body is allowed recovery. In a repetitive, forceful or strenuous work situation, the worker may not be at liberty to take sufficiently long recovery breaks and so runs the risk of overstraining the locomotive structures.

Q2.8) If you observe physical work, pay attention to if the centre of gravity is in the middle of the body or displaced, if the arms are above shoulder level too long or too often, and if the hands are working in the strongest possible position (with a straight, untwisted wrist).

Q2.9) When moving your fingers vigorously or exerting large forces, you should feel the most muscular activity in your underarm. This is where you will feel fatigue if you exert large forces with your fingers, for instance when rock-climbing.

Q2.10)   At the outer extremes of your range of hand flexion/extension, your grip strength should drop considerably – this is why many martial arts will move an opponent's wrist to an extreme flexion when disarming a weapon, such as a knife or baton.

Physical Loading

Q3.1)   Internal loading is caused by the muscular and structural exertion and strain to keep the body in position without adding any additional forces or loading. External loading is caused by lifting, lowering, pushing, pulling, carrying, manipulating materials or tools, etc.

Q3.2.)   Some examples of bad posture induced by the work environment include: bending or twisting to see, stooping or bending to adapt to a work height that is too low, looking down for a long time due to looking at a screen facing upward, cramping arms tightly to the body while working due to a chilly environment, stretching to reach something too far away on in an awkward place, lifting something heavy from the floor with a hunched back, etc.

Q3.3)   Dynamic loading lets a variation of muscles exert force during a task, so that even when they exert a maximal force they promptly get a chance to recover while other muscles work. Static loading loads one or more of the same muscle units repeatedly or constantly, until it/they reach fatigue and the person cannot perform the task for a while.

Q3.4)   Some tips are to seek videos of manual assembly or old war-effort documentary reels from the Second World War to analyse, as many such films feature human workers performing many repeated, strenuous tasks.

Q3.5)   Since this is a self-observation question, try these prompts. Think about your daily habits regarding sleep, exercise and how you transport yourself to and from the different places you spend your working day in. Do you sit still a lot or move around? Do you have work environments that suit your body size? Can you see everything you need? Does your day involve a lot of baggage carrying? Do you load your body symmetrically or asymmetrically?

Anthropometry

Q4.1)   Reasons for choosing a particular anthropometrics database may for example include:
   • wanting measurements for a particular (national) population
   • wanting a recently measured population to take consideration of generational changes
   • specifically seeking out a profession-based sample, e.g. military
   • seeking a database that includes dynamic measurements
   • preferring a database based on body scans rather than manual measurements
   • wanting a combination of national populations to design for an international scope of users

Q4.2.)   This means that anyone whose body segment measurement falls below the population's 5th percentile value, and anyone whose measurement exceeds the population's 95th percentile value, will not find the design appropriate for their body size, but most of the remaining 90% of the population should find it suitable.

Q4.3)   Static body measurements are highly formalized measures of different body segments (defined as the distance between standardized points called landmarks) that are taken with high-precision measurement devices when the person is still. Dynamic measures are measured in motion, often with several locomotive structures engaged to enable maximal reach, movement range, etc.

Q4.4)   Examples of normally distributed body measures in a population include: stature (height), shoe size, upper arm muscle area, sitting height, eye height, etc. These are the measurements that follow a bell curve across a population and have a tendency for one particular measure value to be labelled "average" (when a population can be expected to roughly predict the most common value).

Q4.5)   Examples of non-normally distributed body measures include: hand strength, weight, fat percentage of body weight, etc. These are the measurements that may look skewed in a population, possibly based on lifestyle patterns.

Q4.6)   Almost no person is completely "average" (i.e has the mean or 50th percentile value) in all their body measurements combined, so designing an "average" fixed position solution tailored to that imagined combination of measurements is more likely to be unsuitable for all users. This is because not all body measures are correlated. For example, an "average stature" user might simultaneously have a relatively very long arm span, short legs or wide hips.

This means that it is often best to design to fit a range of sizes. Remember, suitability of fit is only possible with the use of anthropometric measures *relevant to the design problem*, again due to the fact that many body measures do not predict each other.

Q4.7)   A "critical user" has specific body characteristics that constrain the design solution's dimensions; this critical user's measures *must* be considered or the design solution will be a functional failure that does not allow the person to use or operate the solution. For example, the tallest person's height makes them critical when specifying a door height, while a shortest user (or more appropriately, a user with the shortest arm reach, such as a wheelchair user) is critical to the placement of a door handle.

Q4.8)   Look for "one size fits all" solutions; for example, seating heights and widths in public facilities, hand tool sizes in a workshop, shelf and storage heights, window heights, railing diameters, etc.

Q4.9)   This question relates to when the kitchen was built – in some countries, there is a noticeable difference in countertop heights between "generations" of home interiors, and kitchens from the 1930s to '50s for example tended to be tailor-made to the measurements of the woman of the house. However, as demand increased for factory-made home appliances, standard heights became established so that interiors would not conflict with mass-production standard heights of kitchen sinks, retail-purchased appliances like dishwashers, etc. So when compared with modern standard heights for sinks and countertops, it is quite common to find that the older kitchens were lower in height overall.

Cognitive Ergonomics

Q5.1)   Vision, hearing, touch, smell and taste.

Q5.2)   Depending on whether you already knew the answer to the above question and simply recalled the information, *or* if you learned this very recently (within the last few hours) and recalled it, you were in the former case using long-term memory, and in the latter, short-term memory.

Q5.3) See the list of design principles in section 9.5.1 and reflect on how they apply to visual information – such information should support human perceptive abilities (like pattern recognition, sensitivity to size and direction) and correspond to operators' mental models of how the system works.

Particularly relevant design principles are 1, 2, 4, 5, 7, 8, 9, 11, 12, 13.

Q5.4) Attention, perception, memory and mental models.

Q5.5) When a novice performs a task, they need to recall, interpret and process information that is not yet internalized, so their actions are often *knowledge-based*. This could range from explicit problem solving to needing to look up instructions to be sure that correct actions are taken.

When an expert performs a task, they operate quickly, efficiently and in an almost intuitive, automated manner with little delay after a stimulus – i.e., from a *skills* base. This means that their knowledge of how to act appropriately is internalized and reflexive, sometimes to the point where it may be difficult for them to explain how or why they act or react as they do.

Q5.6) Poka yokes are solutions that aim to "error-proof" assembly actions by immediately correcting or alerting operators to mistakes made. For example, a fixture or piece of equipment that makes it impossible to position a part incorrectly functions as a poka yoke. They support cognitive abilities such as visually perceiving if the correct action has been taken, if the right steps or right number of components have been remembered, and by drawing attention to errors.

Q5.7) Look for signs, warning signals, "rules of traffic" where people and vehicles are moving around, uniforms, confirmation signals (like beeps or coloured lights) symbols, repetitive elements, textures in floors and other surfaces, attention-grabbing design solutions, international adaptations, etc.

Q5.8) This reflective exercise may be quite different between individuals depending on preferences, hearing and vision capabilities and ability to filter incoming stimuli, and it may also vary over the course of your life (for example, after a psychosocial illness like burnout or exhaustion, the sensitivity to noise and overstimulation is often more pronounced). It may be a good idea to discuss this question with people of different ages, professions and preferences to understand just how differently we can perceive the "distractiveness" of the same environment.

Psychosocial Factors and Worker Involvement

Q6.1) *Intrinsic motivation* to perform a task arises from an inner sense of meaningfulness for the individual, in such a way that they voluntarily dedicate time and effort to the task.

Extrinsic motivation is fuelled by external reasons separate from the task itself, such as rewards, threats of punishment, a higher purpose or aspects of self-development or recognition.

Q6.2) Positive stress readies the body and mind to take on challenges, by raising alertness and re-distributing nutritional resources to prioritize reaction.

Negative stress appears in situations where the challenge seems unmanageable and the stress hormones lead to a state of discomfort and reduced function of many regenerative processes in the body.

Q6.3)   Chronic stress wears down the body's capabilities for functioning and self-repair, by perpetuat-
        ing elevated levels of hormones that reduce functions of digestion, re-growth, learning and the
        immune system. In other words, chronic stress may contribute to several serious health issues,
        such as an overworked heart, anxiety, muscle tension, digestive problems, high blood pressure,
        exhaustion and weakened capacity to repair and recover. This in turn may greatly reduce a
        person's quality of life.

Q6.4)   In the beginner stage of any learning process, the cognitive challenge of performing even routine
        tasks may be enough of a challenge to keep us alert and engaged (and frequently, skills in routine
        tasks are a necessary stepping-stone to more complex tasks). The risk of committing errors due to
        boredom is low in this stage, because the demands are matched to the worker's available skill level.

Q6.5)   Participatory design is beneficial since it allows the collection of ideas and knowledge from the
        people closest to the work processes (the workers), provides an organized forum for the sharing
        and discussion of such ideas to take place, and by increasing organizational ownership of the final
        solution by offering all stakeholders a say in the proposal.

Q6.6)   Here, the aspects to look for may include activities, technical interfaces and solutions, organiza-
        tional positions and functions, cultural behaviours, etc. in combination. For example, are there
        regular and/or frequent meetings between workers and leaders to re-calibrate the demands, the
        control and the support? Are there systems (online or analogue) to capture, encourage and/or
        reward suggestions, issues, ideas and feedback? Is there an organizational role dedicated to such
        issues? Is there a culture that reinforces frank and open discussions of needed changes?

Data Collection and Task Analysis

Q7.1)   A basic level of good engineering/researcher ethics should ensure that the study or analysis
        includes *informed consent* (participants should be informed about the purpose of the study, what
        is expected of their involvement, and how any collected data will be handled afterwards), treats
        human data confidentially, and if the study is carried out within the context of an organization,
        the engineer/researcher should inform themselves of and follow ethical review requirements for
        that organization.

Q7.2)   An *observation* is non-invasive and aims to influence work operations as little as possible, to gain
        an understanding of how the work normally progresses and what aspects influence the work
        in the as-is situation. Usually, this allows unexpected events to unfold, so as to include as many
        aspects as possible (e.g. organizational and demands).

        An *experiment* involves the implementation of a change to normal operations, for the purpose of
        evaluating the effects of that specific implementation. Here, it is desirable to have as few external
        influencing factors as possible, so it may be that the experimenters "isolate" the object or process
        being studied to control the scope of influencing factors.

Q7.3)   a) This diagram re-joins two tasks that the node above them has been broken down into. This
           means the task breakdown is no longer purely hierarchical, where the level above gives the
           "why" for each subtask.
        b) This diagram does not break down the tasks into several subtasks. Specifying a task into just
           one subtask is not a further breakdown, it is an elaboration, and does not contribute to a hier-
           archical understanding of the task.

Q7.4)   This exercise may turn out differently for different people depending on the interests of the analyst, but a central aspect to reflect on is whether tasks must be performed in any particular sequence. This leads onward to aspects of sequencing and inter-dependencies of process tasks.

Ergonomics Evaluation Methods

Q8.1)   RULA (Rapid Upper Limb Assessment) originated from the textile industry and is appropriate for mostly seated, hand/arm/upper body-intensive work.

REBA (Rapid Entire Body Assessment) originated in hospital/healthcare industry with patient handling and large transient loads as a main focus.

Q8.2)   Most posture-based ergonomics evaluation methods consider just one "frozen" posture at a time, with limited consideration of force- and time-related loading aspects (like how weights are handled, variation and frequency). Many methods also have criteria levels that are tailored to a particular type of work, which may not be universally suitable and therefore results may be met with questioning. Furthermore, some methods' acceptability criteria may be based on a specific population.

Q8.3)   The NIOSH lifting equation's load constant and acceptability criteria come with the disclaimer that the limits are considered safe (under ideal lifting conditions) for 90% of a total working population that is 50% male and 50% female; separated by sex, the weight limits are set to a level that are acceptable for 99% of men and 75% of women.

Q8.4)   Heuristic evaluations are strongly dependent on the ergonomics knowledge and expertise of the analyst to be of any value; it is essential that the "rules of thumb" by which the work system is judged stem from considerable experience and training in ergonomics and human factors. HEs are usually unsystematic, limited in scope and subjective.

Q8.5)   This question requires research into specific standards, but as a first step, we advise our reader to consider the range of parameters that are given acceptability criteria and limits. Are these parameters universally found in all kinds of workplaces, or are some problems specific to particular work sectors?

Q8.6)   This is a "reverse engineering" exercise, where it can provide food for thought to examine the extremes of human motion ranges and how they are interlinked. Changing one body segment from worst possible to best possible while maintaining the others as they were may not even be possible – this provides some insight into how some movements interact with each other to produce an overall risk level.

Q8.7)   This exercise is a useful walkthrough for "sorting" methods into categories of relevance for a particular task at hand or work environment.

Digital Human Modeling

Q9.1)   Reasons for using ergonomic simulation with DHMs include: testing new layouts and processes without exposing real workers to physical risks; enables a proactive approach to workplace design; cheaper than building physical mock-ups for testing; new ideas can be tested for a variety of body shapes, genders, nationalities and ages; lets the designer test many different solutions quickly and cheaply; can be used for training of operators (e.g. to learn a new task); provides visualization of the design proposal that can be used for communication for other stakeholders.

Q9.2)   DHM functionalities that can aid decision making include: visualizations of operator postures while performing tasks in the planned environment; ergonomic analysis tool results that show whether the postures and lifts are acceptable; animations showing a whole movement sequence; path-planning analysis showing collisions; field-of-view, space and reach envelopes showing if all operators can see and reach necessary components.

Q9.3)   Populations are represented by "manikin families", a set of manikins with statistically motivated body measurements that follow the distributions measured for a particular (national) population.

Q9.4)   When observing or analyzing a DHM evaluation case, consider whether the provided visualizations and built-in evaluation results sufficiently communicate the demands of the work tasks and work environment; remember that most DHM programs are a "cleaned up" representation that does not reveal aspects of how well-lit, noisy, dusty, hot/cold or stressful the workplace is or if workers may require protective gear. These aspects may require additional consideration (e.g. using photos of the "real" situation).

Manual Materials Handling

Q10.1)   Risks for quality and safety include: injuries from handling bulky, heavy or hard-to-handle materials; handling of easily scratched or damaged materials increases awkwardness and strain of physical loading; many component variants may result in incorrect selection; high repetition of movements in rearranging or collecting components; bending, stretching and twisting motions may be exacerbated by high repetition; high mental load may result from having many variants, leading to a risk for quality errors.

Q10.2)   Pros of kitting include: better quality and flexibility, less time spent walking and collecting (for the assembler), less time variation in completing assembly tasks, aids operator learning regarding product components, better materials control (components are at hand), better visibility of the shop floor and assembly line flow, pace keeper (takt time), operators can focus on assembly rather than selecting components.

Cons include: may require additional staff to carry out separately, may take up more space in total since the original packaging is separate from the materials façade, limited to small or medium-sized components due to size limitations of generic containers.

Q10.3)   Pros of line stocking include: original packaging is delivered straight to the assembly without extra work steps; stock is continually available, so no extra storage space is needed; easier to automate delivery of stock; easy to select a new component if defective ones are found.

Cons include: capital is tied up in stock; shop floor space is taken up with large pallets and containers; may result in lack of space at workstations and passageways, especially if many variants are produced; requires a lot of time for walking, removing packaging and selecting components (in turn requiring a higher cognitive load).

Q10.4)   Pros of small containers include: lower peak loading on back and shoulders compared to large pallets; requires shorter supply racks, reducing the space needed to display components; reduces amount of walking time, by allowing different components to be stored closer together; greater flexibility and allows more product variants.

Cons include: Requires frequent replenishment, meaning more "traffic"; requires additional work with specific kitting and/or engagement of dedicated kitting staff.

Q10.5)   Here, consider examples where it is obvious that regular replenishment of a certain good or material is necessary for operations to keep on.

   In retail, consider shops where the stock is readily transferred to within customer reach, such as in supermarkets or building materials/home improvement shops. How are the goods labelled, positioned and transported?

Q10.6)   Here, observe e.g. sales of "kits" of materials for a specific purpose or project, such as when the right amounts of material, tools and finishing products (e.g. paint or lacquer) to something specific, like a shelf or storage shed, or when a lot of different goods types related to a specific type of project are found together in the store.

The Economics of Ergonomics

Q11.1)   See the list in section 11.1.1. Individual costs of poor ergonomics tend to be counted from the point where the worker's ability to carry out work is negatively impacted.

Q11.2)   "Hidden" costs of sick leave come as ripple effects from the worker's absence. These include costs incurred due to recruiting and training replacement staff, reduced productivity and quality while new staff are trained, slower production speed, legal compensation costs, etc.

Q11.3)   From an industrial perspective, poor assembly ergonomics can result in more quality errors (leading to more scrapped material, re-work and increased lead time), more warehousing space needed for re-work, more staff time requirements and poor reputation as a workplace that can ward off future potential workers or even customers.

Q11.4)   Examples of expected gains from investing in better ergonomics may include: fewer production errors due to poor reach, strength transfer or other performance impairment; shorter assembly time due to elimination of unnecessary or corrective motions; better accessibility and increased functionality for a wider range of workers; reduced levels of exposure to hazards; (from a systemic point of view) better goodwill among workers, leading to loyalty and building of competence.

Q11.5)   As a general remark when comparing ergonomics/productivity case studies, it is worthwhile noticing how the gains are described and over what kind of time perspective. Sometimes the gains are indirectly related to economic results, rather than given in bottom-line savings. This is the case particularly when improvements are studied shortly after an intervention, the presumed gains are often described in terms of eliminated injury risks, faster processes or better quality per "produced unit" – a direct cost comparison is usually only meaningful to report when the intervention has been in place for some time, for example the projected time for regaining the investment cost.

Q11.6)   This relates to identifying the systems view in the case study – were only technological or direct injury-related aspects included in the solution and expected impacts, or were there mentions of impacts also on intangible aspects like worker behaviour, lessened scrap, productivity improvements and customer impressions? A wider systems view may lead to more potential ways to account for the worth of an ergonomics improvement.

Work Environmental Factors

Q12.1)   Extreme heat   Extreme cold

- Painful cramps   - Apathy
- Stomach trouble (intestinal impairment)   - Disorientation
- Failure of the body to regulate its   - Shallow breathing
  temperature, leading to overheating   - Failure of the body to regulate its
  (hyperthermia)   temperature (hypothermia)
- Performance impairment and errors,   - Frostbite – leading to tissue damage
  accidents

Q12.2)   Full-body vibration may (in the short term) lead to nausea, difficulty to see, numbness, fatigue
          in the muscles, difficulty with precision movements, excessive strain on joints, stomach trouble
          and headaches, as well as (in the long term) circulatory, bowel, respiratory and low-back
          disorders.

Q12.3)   Elimination of the noise source, insulation of the noise source, regulation using standards, and
          hearing protection devices.

Q12.4)   Irrelevant high-intensity light that does not contribute to better illumination, but irritates and
          overwhelms our sense of vision, leading to temporary inability to see.

Q12.5)   An effective alarm signal must:
- be identifiable and detectable – i.e. easy to differentiate and notice from regular ambient noise,
  in terms of frequency and loudness
- be at a frequency that is not too high-pitched, as workers with age-related hearing loss might
  not be able to perceive it easily
- be at a loudness level that attracts attention but does not cause extreme pain

Q12.6)   Ionizing radiation is considered very damaging to the human body structures, since it can
          detach electrons from atoms in human tissues, and mostly cannot be turned off.

          Non-ionizing radiation is not as dangerous, and a source of such radiation can generally be
          turned off.

Q12.7)   Clothing and protective gear against extreme temperatures may have the following disadvantages:
- The human loses the direct sense of touch towards the environment, which implies that this
  sense may miss important information.
- Thick or bulky insulation layers may decrease the precision of hand movements.
- Some protective gear may impair sight and hearing as well as touch.
- In many human-machine interfaces, controls may rely on tactile ability and skin conductance –
  protective gear makes it difficult to manipulate small controls (dimensioned for unprotected
  fingers and hands) or touch-screens.

Q12.8)   Try searching online for the equipment names or measurement units named in the different
          sections, coupled with "work environment" or similar keywords. You will find that there are
          many purveyors of equipment of many different sophistication levels – try to see if any that
          you find seem to be tailored to a specific type of industrial work environment, for example by
          searching for customer cases!

Q12.9)   This question is more of a guide for any "first visit" to a workplace. Often, the operations of the work themselves may cause some work-environmental disadvantages (such as carpentry and machining creating dust particles or noise), and the design of a healthy work environment must take into account that any industrial hygiene or protection measures are only likely to be effective if they do not hinder the on-going work from being done.

Social Sustainability

Q13.1)   The WCED defines *Sustainable Development* as "Development that meets the needs of the present without compromising the ability of future generations to meet their own needs."

Q13.2)   Corporate Social Responsibility, loosely defined, it is a commitment by corporations to have a beneficial impact on the humans and environments they interact with.

Q13.3)   1) Fewer young people being born (i.e. a smaller talent pool in the future)
         2) Elderly population (physical condition and performance may deteriorate, but their experience and skill may be hard to transfer)
         3) Skills gap (many people are not equipped with the education and skills to handle future manufacturing demands)
         4) Unclear needs of the future workforce (due to their diversity and variation in skills, personal goals and values)

         (There are more, but the above are the ones described in this book.)

Q13.4)   In relation to retaining staff, it is likely that recurring training will give them an impression that their company cares for their personal development and competence. In the best case, this may attract the staff to stay and develop loyalty toward the company (if training is systematic and systemically implemented and either internally rewarding or coupled to a rewards structure).

Q13.5)   Inclusive workplace design makes it possible for a wider variety of people to perform value-adding work in a workplace, thus enabling a greater proportion of a societal population to be able to gain employment and be productive no matter what their physical, mental or social limitations.

Q13.6)   Some examples of what employers can do (note: this is not an exhaustive list, and the answers are intentionally broad!):

         Demographic change:
         Attract, recruit and retain a wider variety of future workers with regard to age, gender, skill range and background, and make the workplace functional for all their needs.

         Empowerment, participation and access:
         Encourage and systematize ways for employees (and perhaps even customers and suppliers) to engage in how operations and business is run.

         Identity, sense of place and culture:
         Cultivate an inclusive organizational identity that employees can agree with, contribute to and feel pride in belonging to. This may tie into issues of ethics.

Health and safety:
Value health and safety highly. Follow the design guidelines in this book and distribute the responsibility for these aspects throughout the organization.

Social mixing and cohesion:
Ensure that groups within the organization intermingle, exchange ideas and engage with each other as a community across organizational and professional/disciplinary boundaries.

Social capital:
Ensure that communication, networks and trust within groups in the organization are healthy and functional.

Well-being, happiness and quality of life:
Offer time, facilities and means for employees to pursue a healthy and balanced life-style. Implement policies and organizational support to maintain a family life, social bonds and a personally meaningful lifestyle.

Q13.7)   Here, we leave it up to our reader to find a contemporary model organization – because whatever we write here at the time this book is published, at any time changes may come about that can place a superstar company in controversy. But seeking coherence between company culture, future vision and employee incentives may be key to finding a sustainably attractive employer.

www.ingramcontent.com/pod-product-compliance
Lightning Source LLC
Chambersburg PA
CBHW042032220326
41598CB00074BA/7409